Once Upon a Time in Facilities Management

Once Upon a Time in Facilities Management

Tales from an Organizational Netherworld

Patricia McCarroll
John Hassard

OXFORD
UNIVERSITY PRESS

OXFORD
UNIVERSITY PRESS

Great Clarendon Street, Oxford, OX2 6DP,
United Kingdom

Oxford University Press is a department of the University of Oxford.
It furthers the University's objective of excellence in research, scholarship,
and education by publishing worldwide. Oxford is a registered trade mark of
Oxford University Press in the UK and in certain other countries

Published in the United States of America by Oxford University Press
198 Madison Avenue, New York, NY 10016, United States of America

British Library Cataloguing in Publication Data
Data available

Library of Congress Control Number: 2024932977

ISBN 9780192848192

DOI: 10.1093/9780191943492.001.0001

Printed and bound by
CPI Group (UK) Ltd, Croydon, CR0 4YY

Preface

What would the world of work look like if interpreted through the lens of the fairytale? To answer this question, we explore storied spaces and metaphorical archetypes in the study of business, management, and organization. Taking inspiration from Gareth Morgan's exploration of the concept of metaphor—in his book *Images of Organization* (1986)—our work reveals representations of business and management activities through similar analogical means, in this case through taking recourse to fairytales, folklore, and fables in sociological exploration.

The book, however, is not exclusively theoretical in orientation, for it has an equally important empirical focus—exploring a business service rarely studied in the management literature, facilities management (or FM), a 'secondary service' of non-core and increasing outsourced organizational functions. An in-depth experiential appreciation of FM is provided, one that assesses the people, practices, and processes of the service in a study which also highlights the characteristic liminality or marginality of the sector's professional activities. Emphasis is placed on illuminating the storytelling nature of the service, using primarily the genre of fairytales to identify representational archetypes (e.g. queen, shadow, sage, trickster, adventurer, eternal child) within FM's storied spaces. In the process, three central professional characters (essentially mechanisms of delivery) are identified—the professional consultant, the external service provider, and the in-house function—with these forming the operational foci of fairytales explaining the culture and symbolism of FM as a business service.

Finally, in terms of research findings, this innovative methodological approach reveals much perceived illegitimacy within the FM function itself, albeit a situation which is seen to facilitate the strategic metamorphosis increasingly required of secondary services in uncertain and dynamic organizational settings. Although noting how FM is often misunderstood outside of professional practice, the storied spaces of the three delivery forms reveal no such ambiguity, for instead a process of recognized connection among them is found: this reflecting awareness of their archetypally 'shadowy' role as the 'necessary evil' in the organizational landscape. The book concludes by extrapolating findings from the investigation to inform a discussion of the contributions of folkloric analysis to organization theory and our understanding of business and management practice more widely.

Contents

List of Figures

List of Tables

1
Once Upon a Time

> Facilities managers? They are wee fairy folk, who come out at night when everyone else is sleeping, and fix everything before we come back to work in the morning.
>
> **Extract from a presentation by comedian Dara O'Briain to British Institute of Facilities Management [BIFM] Annual Excellence Awards, 2007**

As the above quote from Dara O'Briain implies, the identity of secondary services within organizations is a complex issue, especially given that fragmented boundaries and rapid change are hallmarks of many contemporary companies (Alexander and Price 2012, Bleistein 2017, Moravec 2013). The research presented in this book seeks to understand the world of such a secondary service, facilities management (FM), from the viewpoint of the profession's practitioners. Taking an allegorical approach, folkloric and organizational storytelling frameworks are developed and entwined to form an interdisciplinary method directed at producing what is ultimately labelled *artificial folklore*. This approach synthesizes organizational fairytales methodologically but remains true to an analytical tradition that recognizes 'the changeability of folklore as compared to the stability of literature' (Benson 2003: 20).

The study of FM within this landscape offers insight into how the securing of a sense of disciplinal legitimacy is achieved, offering in the process the opportunity to define emergent practices, knowledge, and relations at a time when organizational forms themselves are indistinct, uncertain, and in flux (Hanelt et al. 2021, Volberda et al. 2021). As such, the study of FM represents not only a vehicle to understand an unfamiliar structural practice, but also the prospect of identifying new social processes of relevance for comprehending contemporary forms of organizing, and, in this case, specifically opportunities for 'exploring and enacting symbolic practices which then become embodied in future organizational arrangements' (Lucas 2014: 2).

Once Upon a Time in Facilities Management. Patricia McCarroll and John Hassard, Oxford University Press.
 DOI: 10.1093/9780191943492.003.0001

This extends not only to present but also future iterations of FM, and particularly in terms of the ways FM reacts to the needs and demands of the organization it is affiliated with. At the level of practice, this study answers a call for a deeper understanding of the present state of the FM discipline, function, and profession (Tauscher and Clayton 2022). Additionally, it facilitates the introduction of conceptual and theoretical frameworks that draw on knowledge and understanding from a range of academic fields—but in particular from anthropology, folklore, and sociology.

Moreover, it is recognized that FM, as an organizational practice, is following a path already laid by other secondary corporate disciplines, such as information technology (IT), human resources management (HRM), and more recently, project management and hospitality management. Such disciplines have created or sustained a set of firm identities and recognized functions but also faced the risk of being sub-contracted or outsourced As such, in many respects the study of FM offers insights into shared experiences and symbolic understandings affecting many other functions and disciplines within the realm of corporate organization.

The FM industry in the United Kingdom is also unique, due to the long history of debate concerning what FM actually is, and what it should be, leading to a complex and convoluted identity (Leaman 1992). The generic claim often made within textbooks is that FM primarily concerns the management of 'people', 'process', and 'place' (Price and Akhlaghi 1999). However, it is also claimed, more specifically, that the main role of FM is to manage the 'built environment', through various 'operational' roles (Tucker and Masuri 2018). Traditionally, this has been manifested in the work of janitors, porters, caterers, receptionists, cleaners, helpdesk operators, maintenance teams, and estate managers (Becker 1990; see also Atkin and Brooks 2021).

This, in turn, has led to an ambiguous organizational purpose (Drion et al. 2012), and to a discipline that is perpetually striving to 'come of age' (Alexander 2003, Lowry 2017). In this regard, FM is an idiosyncratic entity to study as it fights to legitimize its function and identity, whilst being part of a wider organizational landscape which is increasingly negotiated, fluid, and temporal. This is succinctly described by Price (2003: 13) who suggests: 'FM is firstly engaged in a Darwinian struggle to secure its niche in a wider business "ecology" and secondly, it is a system in which competing interpretations, and the institutions that hold them, are themselves engaged in such a struggle.'

FM therefore holds a precarious functional position in terms of being unrecognized and hidden within the structural setting—an organizational netherworld—positioning it as the 'alternative voice of organization; that is, voices other than those privileged by the organization itself' (Höpfl 2000: 15). Paradoxically, FM is highly visible as it represents the physical embodiment of organization in its role of maintaining the built environment and the many social, technical, and material factors within it. FM then is situated as an organizational discipline that affects the realities of others within places of work by maintaining the physical infrastructure (Küpers 2011).

In this study, the medium of folklore is used both as a means to capture a picture of FM within its current state and to aid understanding of the corporate domain of secondary services generally. From a broad perspective, folklore is a cultural practice that reflects the traditions of the past but also with a view to forecasting the future (Blank and Howard 2013, Goulding et al. 2018), so that stages of evolution are captured within its stories. The use of folklore within this research is advanced with the understanding that contemporary organizations are in themselves on an evolutionary journey. As Marmot and Eley (2000: 4) suggested in this regard, 'a new generation of organizational patterns and social consequences is upon us', one in which 'offices have flowered as a building form, and a social structure, for a fairly short, intense period. They seem to have totally permeated our understanding of the world of work. But they could be a short-lived phenomenon, a transitional stage in an economic evolution.' The use of folklore offers a unique way to gain further insight into the world of organizations, just in the way that the Grimm Brothers' infamous collections captured the social, cultural, and political issues of their times.

We claim further that FM is arguably the 'storyteller' of the business, shaping the demands and designs of 'folk' most visibly through FM's role in managing the built environment, which is 'not so much a product as a process and a constantly unfolding narrative' (Alexander and Price 2012: 1). This study aims to capture the stories of the storytellers themselves, with the understanding that these tales represent fragments of folktales within the industry—the folk here being FM practitioners. Folklore in this sense is contemporary folklore and used analytically in connection to concepts from academic fields such as sociology, psychology, and cultural anthropology. It specifically concentrates on the subset of fairy stories (Baruch 2009) as a genre of folktales (Holbek 1987). Defined as stories that contain 'fantastic elements' (Trubshaw 2002), fairytales reflect both unique and universal imagery (Dégh 1994). Such imagery, however, remains an area of analysis not yet fully investigated within a wider business and management context. For organizational analysis, fairytales provide an emotional yet familiar form of sociological connection; one renowned for simplifying complex social, technical, and material structures and processes.

1.1 Analytical Approach

This study is set in the context of a professional identity crisis within FM. On one level, FM embodies a discipline of strategic importance (Lowry 2017, Nutt 2000), whilst on another it symbolizes 'organizational hyperbole as a management fad' (McCarroll 2017: 219). This ongoing debate is one that the first author, in particular, is familiar with on several dimensions. As an FM practitioner for over a decade, the range of roles, responsibilities, and expectations of service delivery (and the associated confusion therewith) was experienced first hand. This spanned across four different organizations, three based in the private sector and one within a local authority. The different FM delivery mechanisms were also experienced directly by the first author: through working professionally as both an external service provider and as an in-house facilities manager. Experience of the world of FM then continued, but in a different domain—as (after completing post-graduate work) the first author became a university lecturer in FM, at which point she embarked on research on business services in collaboration with the second author, a long-time researcher on management and organization. This practical experience on the part of the first author has allowed for not only the durability of the FM 'debate' to be appreciated, but also the scale, notably as communicated through the stories of the facilities managers who have represented the majority of students she has taught.

This book aims to act therefore as a conduit to understanding the confusion and irrationality that increasingly complex organizational relationships can bring to a secondary service such as FM. Where the discipline of FM aims to manage the interactions between people, the work they do, and the work environment, this research questions that reality, seeking instead to understand such relations by focusing on the perceptions of the main 'character': the FM practitioner. In this case, it leads to an investigation into the current nature and role of FM in the United Kingdom. To do so, mechanisms utilized within folklore are applied to this organizational analysis, effectively creating *artificial folklore*. This sees the taking of a mainstream business issue—such as organizational identity—adjusting it for application in a new business discipline, and then utilizing new data to seek further understanding of the phenomenon. In so doing, the approach developed aims to consider longstanding tropes within new contexts, in this case, folklore in unfamiliar business and organizational domains. Philosophically, the method borrows inspiration from the novelist and journalist Angela Carter (1983: 69) and her view that 'most intellectual development depends upon new readings of old texts', with the accompanying suggestion that 'I am all for putting new wine in old bottles, especially if the pressure of the new wine makes the bottles explode.' This is arguably the essence of folklore (Blank and Howard 2013) as it continues to reflect shared traditions and communal pasts and futures. Fairytales expose elements that are largely unconscious, uncovering behaviours that may reinforce wider institutional and institutionalized behaviours. Their use within this research is to act as a framework to make such organizational dynamics explicit.

The use of a fairytale framework will also include, purposively, analysis of the role of absurdity and irrationality in organizational affairs, uncovering the basic assumptions held about FM as told by the practitioners themselves. Basic conventions in this context 'represent a system of shared meaning that governs collective perceptions, thoughts, feelings, and actions. This system forms the implicit, "taken-for-granted" worldview that defines the organization's concept of itself, its constituents and its environment' (Schneider and Shrivastaval 1988: 494). The intention is to increase understanding of the complex social—primarily organization-based—situations facing FM practitioners by capturing the meanings that actors create and inscribe within a projected storyline, thus allowing a window into roles, identities, and beliefs within the field. The objective, therefore, is not to pigeonhole FM organizationally, but instead to comprehend the uncertainty operating within its functional and professional boundaries.

The question over what FM represents has long been debated (McCarroll 2017), but in this inquiry the issue is redefined in that facilities managers themselves are asked to tell their own 'story of FM', from their experiences of working in the sector. The data therefore comes from practitioners' perceptions of the characters they play and the story plots they experience. This forms an understanding of the 'lifeworld of organizational actors' (Styhre 2017: 16) employed within modern FM. To support this, the methodology of *artificial folklore* has been developed, with this, as noted, being underscored by influences from a wide range of academic disciplines, such as anthropology, folklore, philosophy, psychology, and sociology.

We must also recognize that 'artificial' elements have been introduced to capture these stories. However, these are considered as artificial rather than false or bogus (Dorson 1963), as it is proposed that these stories are representative of longstanding tales, and that the use of the fairytale framework is a means of capturing accounts at a particular moment in time. The authenticity of the tale remains: it has not been extracted from its context, but instead used as a tool to further our organizational understanding (Zipes 2005). In so doing, this innovative arts-based research study has three main aims.

1.1.1 Aim One: To Unlock Narrative Patterns within the FM Storied Space

Storytelling is recognized as a cognitive device that is a fundamental skill, enabling complex and disjointed experiences to be understood and acted upon (Gabriel 2000). This can be assumed as the creation of a series of social ecosystems, identified as 'storied spaces' (Brewer and Dourish 2008). Storied spaces exist at individual, group, professional, and national levels, for example, and behaviour is therefore adjusted accordingly to the norm accepted in each state (Boje and Baskin 2011). By treating the FM profession as a storied space, it will allow the accepted behaviours associated with FM to be illustrated, as they are formed by the interactions of the collective within this space.

The medium of fairytales will be used to investigate the FM storied space, as this genre of folklore concentrates on the reoccurring motifs and plots that form traditions, with these manifesting as the patterns of storylines. The aim here is to identify the reoccurring plots and themes that FM practitioners believe exist within their storied space, capturing the storylines they create that represent the norms of FM and help understand the identity and fluidity of the practice. By using fairytales as an analogy to explore the storied space, it will capture these reoccurring patterns and therefore uncover the realities held by FM practitioners, allowing the research to 'tune into operative dynamics that would otherwise remain very covert and inaccessible' (Smith and Simmons 1983: 377).

1.1.2 Aim Two: To Uncover the Metaphorical Archetypes Held within FM

With research aim one highlighting the accepted plots and therefore organizational realities of FM, research aim two is then to identify the different archetypes held within the industry. By maintaining the metaphorical approach of the fairytale, this will highlight what roles facilities managers cast themselves and others into, in line with the understanding that an employee's identity affects how they undertake their role (Butler et al. 2014).

These are the learned behaviours of the discipline—the roles played either voluntarily or as pre-determined by other forces via the generic story characters that FM emotionally resonates with. These archetypes reflect an 'operative paradigm . . . in which an individual can experience the world, be compelled to action, and provide a model for behaviour' (Shadraconis 2013: 1).

In this way the archetypes, as held within FM, can provide a framework to support the storied space.

1.1.3 Aim Three: To Create an *Artificial Folklore* Framework for Organizational Research

This aim corresponds with the phantasmagorical approach to creating fairy stories that are allegories of FM, by developing an *artificial folklore* framework. The aim is not to create a factual story relaying events that occurred, but the creation of authentic stories that represent the lifeworld of contemporary practitioners of FM. This is the creation of a new way to perceive and interpret the discipline—one generating vivid experiences that might otherwise go unnoticed. As such, the research answers the call to 'stop treating the social and the material as distinct and largely independent spheres for organizational life' (van Marrewijk and Yanow, 2010: 41).

1.2 Chapter Overviews

There are nine chapters overall. This initial chapter introduces the research, describing the practice of FM and positioning this function as an example of a secondary business service, given its notionally 'hidden' position within the organizational landscape. The approach to organizational storytelling is outlined, and three research aims are defined to guide the reader through the study. As noted, the first is unlocking narrative patterns within FM's storied space by applying a fairytale approach, which identifies reoccurring plots and themes that reflect the norms and identities associated with FM work. The second is to uncover metaphorical archetypes, highlighting the roles facilities managers cast themselves and others into in order to understand how an employee's identity affects the undertaking of their role. Finally, we introduce the new research method of artificial folklore, where the intention is not to create a factual story relaying events that occurred, but to generate interpretive stories that imaginatively explain contemporary practices in FM.

Chapter 2 defines FM from academic, corporate, and institutional sources, commencing with a chronological tracking of the discipline's evolution. Various service dimensions are identified, allowing comparison of professional attributes in assessing the validity or otherwise of characterizing FM as a standalone discipline. Subsequently, we assess what unites FM practitioners in the sense of belonging to a recognizable professional community.

Here the association of FM with the built environment is detailed alongside analysis of contemporary knowledge work and associated spatio-temporal changes. In so doing, the concept of liminality is introduced to emphasize the various negotiated and shifting organizational boundaries of FM. The final section concentrates on FM's delivery mechanisms, highlighting the growing association of the discipline with outsourcing. This identifies moves to an increasingly intangible service—a discussion leading to consideration of storytelling as a means to communicate wider issues of organizational culture, meaning, and change.

The research approach detailed in Chapter 3 concentrates on developing organizational storytelling in a way that is methodologically related to folklore. Here the concept of a storied space is investigated as one which interweaves both fiction and fact, and notably through analysis driven by the vehicle of the fairytale. In so doing, a methodology of organizational storytelling is developed in a way that utilizes the potential analytical roles of fiction, symbolism, and allegory. This provides the basis for a further discussion—the specific use of folklore and fairytales as the basis for formal research in business, management, and organization studies. The narrative structure of the fairytale is examined, with this suggesting a method for allowing story arcs to be analysed and archetypes identified in line with key concepts for the investigation, such as liminality and metamorphosis.

Earlier chapters outlined how the study seeks to unlock narrative patterns within the FM storied space and in turn uncover FM archetypes. Chapter 4 concentrates on the data collection process, which combines fact-based information with art-based methods through the introduction and use of the fairytale-based research framework generated previously. This research process is directed ultimately at forming an 'artificial folklore' methodology for organizational research and analysis. By doing so and positioning FM practitioners as the storytellers of their own practice, the methodology elicits novel images and evokes new associations in terms of how FM is experienced, explained, and defined. Specifically, this creates novel understanding through factual recollections being aligned with fictional 'casting', as the chapter details how research materials generated from the three-pronged data collection process—deconstruction, collaboration, interviews—were produced, with each analytical phase described in turn.

Chapter 5 details the development of the empirical stages of the research. Here the operational focus for analysis centres on the three main FM service delivery mechanisms defined—the professional consultant, the external service provider, and the in-house function. In research terms, initially the tale of *Rumpelstiltskin* forms the basis for a pilot deconstruction analysis—one facilitating a clear casting across the four main characters of the story. Second, the main issues raised from a planned collaborative research approach to storytelling are described before, thirdly, the more detailed interview-based analysis—comprising the gleaning of individual lived-experience stories and forms of creative scenario re-imagining—are outlined. This final approach allowed identity narratives to be assessed in line with the three dominant fairytales emerging in this phase of the work—the consultants' *Hansel and Gretel*, the service providers' *Goldilocks*, and the in-house function's *Cinderella*.

In explaining organizational identity and meaning, the forms of information generated by the new folkloric approach to interpretive analysis suggest that each of the FM service streams (professional consultant, external service provider, and in-house function) favours a different principal illustrative fairytale. To interpret the tales in an applied sense, analysis of each is provided in Chapter 6, in terms of work-related issues of experience, relationships, and change. Specifically, the three primary tales selected (*Hansel and Gretel*, *Goldilocks*, and *Cinderella*) are retold by a narrative process that takes reference to the 'Ladybird Classics' series as a guide to depicting subjects, issues, and themes. This ultimately facilitates a detailed organizational analysis of each fairytale, encompassing how they both surfaced within the current FM literature and how interpretations of them were made through empirical inquiry.

The focus of Chapter 7 is the development of a comparative approach whereby the chosen fairytale from each delivery steam is deployed as a consistent narrative structure from which to extrapolate, interpret, and evaluate the projected organizational perceptions of FM practitioners. Themes such as role fluidity, strategic metamorphosis, structural liminality, relationship realignment, and corporate enchantment habitually undergirded the comparative analysis, which was driven principally by lived-experiences stories, there being substantial overlap in interpretation among streams in many cases. This allowed established patterns of FM storied spaces and archetypes, identified across the delivery mechanisms, to form platforms underpinning the final authoring of a ghostwritten fairytale (in Chapter 8)—a process which seeks to capture the idiosyncratic characterizations that constitute this under-researched organizational practice. These detailed stages of analysis were repeated for each of the three delivery divisions, ensuring a consistent comparative approach.

Building on previous stages of analysis, Chapter 8 concerns the final development of method building from our interpretation of data in this narrative account. Here we highlight patterns of FM's storied spaces and focus on the plot and character castings of story-forming, in a process that generates a collective, artificial fairytale 'from FM, about FM'. This sees key findings identified and presented in support of research aims one and two. Reflecting on the research process, we overview how the roles and structures of fairytales have been used to frame the various storied spaces of FM, defining them and deploying fiction both to grant meaning to and identify collective symbols from individual stories. To this end, this chapter on developing the artificial folklore of FM culminates with a ghostwritten tale from each of the three service divisions, thus completing the novel methodological approach originally outlined in research aim three.

This research has analysed the secondary service of FM as an example of a contemporary non-core business function developing the method of artificial folklore as an analytical approach to explain its modes of work and organization. The multi-disciplinary approach developed here has expanded the idea that fairytales can be revised, retold, re-imagined, subverted, and invented as they respond to changes, for example, in social values and corporate needs over time, with the final product being their application to the analysis of a contemporary organizational setting. The ninth and final chapter details, therefore, the main contributions of the study to both organizational knowledge and professional practice, as well as suggesting recommendations for further research in this area of interpretive method building.

2
The Tale of Facilities Management

This chapter forms the first literature-based review in support of the research aims, analysing the current state of FM to help define it from academic, corporate, and institutional sources. It develops in three sections, with the initial analysis (Section 2.1) offering a chronological tracking of the discipline. This allows several important organizational dimensions to be recognized—such as 'hard' (physical) versus 'soft' (people) forms of FM, and 'core' (strategic) versus 'non-core' (operational) functions. Different delivery mechanisms are identified, allowing a comparison of relative professional attributes to be made and examined. With such a concentration in the literature on what FM is and is not—and thus awareness of key specialized differences within the discipline—the discussion also considers what constitutes a uniting sense of belonging in terms of a professional community.

The association with the built environment is explored in Section 2.2, identifying the rise of knowledge workers and concepts such as the 'third place' of contemporary working (Jeffres et al. 2009). This identifies a possible pressure on FM to refocus its practical role away from work in traditional office spaces, which are increasingly forecast to disappear. Here, the concept of liminality is introduced to emphasize the negotiated spatial and temporal boundaries of FM (Hassard and Morris 2020). This is not only in its role of controlling the physical environment, but also in terms of how this state illustrates the increasingly precarious nature of FM—as an instigator of a liminal state and as an 'outside' department disturbing the equilibrium (e.g. through introducing hot-desking); FM being within a liminal state when viewed as a separate entity to the rest of the organization.

The final section (Section 2.3) concentrates on the delivery mechanisms of FM, addressing the growing association with outsourcing. This further supports the presence of a liminal state—allowing a focus on the role of FM to create a new sense of belonging within the discipline—but also analysing the sense of disciplinary belonging held within client organizations. This move to an increasingly intangible service aligns with the contractual nature of FM, leading to a discussion of the role of storytelling to communicate wider organizational meaning, one incorporating areas of enchantment, spirituality,

Once Upon a Time in Facilities Management. Patricia McCarroll and John Hassard, Oxford University Press.
 DOI: 10.1093/9780191943492.003.0002

and meaning. This leads to consideration of storytelling as a competency that may aid and give legitimacy to a profession that has existed without a stable identity for half a century.

2.1 FM: The Birth of an Organizational Character

FM is a broad discipline, one increasingly associating itself with more managerial than facility practices professionally. Academically, it is supported by three main journals. The most enduring, and only officially ranked, journal (one star in 2021 Chartered Association of Business Schools rankings), *Facilities*, has been published since 1983, being developed from an original in-house periodical of Frank Duffy's architectural firm DEGW (Price 2003). This was joined in 2002 by the *Journal of Facilities Management* and then in 2010 by the *International Journal of Facility Management*. The latter journal represents a more global approach to the discipline through its association with IFMA (International Facility Management Association), based in the United States.

The term facilities management (FM) was being used in the early 1970s in loose reference to the data-processing outsourcing strategies that the banking industry was embarking on at the time. By the 1980s, the term expanded to include reference to the built environment. Now FM started to become recognized as a discipline (Steenhuizen et al. 2014, Price 2001), and in particular in response to the increasingly complex nature of building management and performance (Leaman 1992, Finch 1992). Factors such as global competition, the high cost of space, and increasing employee expectations are all considered influences that stimulated the birth of an organized FM function at this time.

An influential catalyst in the creation of FM was the proliferation of information technology (IT) as a secondary service within the workplace. Where this practice evolved rapidly to form an established discipline, FM in comparison was still considered as 'emerging', causing many practitioners to campaign for greater attention, identity, and respect for the discipline (Thomson 1990, Tauscher and Clayton 2022). As Price and Akhlaghi (1999: 162) argued: 'If building management represents the first generation of FM, and the integration of people, process and places the second generation, then the third generation FM might be seen as more concerned with the creation of space which enable different levels and forms of performance.'

Within the United Kingdom, FM is represented through the professional body, the Institute of Workplace and Facilities Management (IWFM), which was founded in 1993. Currently supporting around fifteen thousand registered members, it defines FM as the 'organizational function which integrates people, place and process within the built environment with the purpose of improving the quality of life of people and the productivity of the core business' (https://www.iwfm.org.uk/what-is-facilities-management). However, a strong operational definition is apparent, as the IWFM report goes on to list the common associated FM roles as being 'catering, cleaning, building maintenance, environmental services, security and reception' (https://www.iwfm.org.uk/what-is-facilities-management).

The main declared definition of IWFM corresponds with that of the larger, international, IFMA (International Facility Management Association), which declares FM as 'a profession that encompasses multiple disciplines to ensure functionality of the built environment by integrating people, place, process and technology' (https://www.ifma.org/about/what-is-facility-management). Founded in 1980, the IFMA is the largest international professional body of FM, although it is predominately based within the United States, with most of its chapters located there. FM is thus well established within the United States, not appearing to have stagnated in terms of the definition of its operations in the way arguably the United Kingdom has. A possible reason for this is indicated in the name difference—Facilities Management in the United Kingdom, but Facility Management in the United States. In terms of this research, however, facilities management and facility management are used interchangeably.

Nevertheless, the key elements of 'people', 'process', and 'place', as stated by the two main industry bodies, directly correspond to much of the current research within UK FM. Given that property and people are often the largest expenses in a business or corporation (Price 2003), the positioning of FM as the process element that acts as the hinge between them illustrates its potential impact within organizational settings. This not only reflects the organization's ability to reduce costs operationally, but also to add value strategically.

These definitions, however, are arguably vague in nature and often have not been widely adopted, internally or externally, as a way of defining FM. Therefore, it could be argued that FM is barely understood in terms of presenting a recognized function within business. Indeed, a study in Switzerland found that 79 per cent of the public had never heard of the term 'facilities management' (Coenen et al. 2010), a statistic which could possibly imply a similar level of awareness might exist within the UK public.

From within the discipline, Tay and Ooi (2001: 357) argued 'the profession still suffers from an acute identity crisis', especially given 'the roles and scope of duties of one facilities manager may be vastly different from another'. Externally, authors such as Leaman (1992) have argued that FM is not a credible profession, as there is a lack of any generalized agreement as to what FM actually is (Drion et al. 2012). With no recognized, legitimate function, it is reasoned that FM has diluted its market position to the extent that it is regarded as a minor non-core service (Becker 1990). It has even been argued that the emergence of FM as a discipline may have been dysfunctional to the extent that it may have distracted industry from its core performance goals. As Thomson (1990: 12) argued: 'Facilities Management in the UK has failed because it has missed the most central issues and immersed itself in complicated explanation and lists of activities that appear to encourage still more diverse skills into the fold and create more confusion.'

However, instead of viewing this seeming confusion over identity as detrimental to the discipline, it could also be considered as the basis of a counterargument. It is argued that regardless of the progress made with regards to the definition of FM and its understanding in business, the definition is not fixed (Royal Institute of Chartered Surveyors [RICS] 2013). It will always vary in line with the nature of the primary industry it is supporting as a secondary service. In fact, Waheed and Fernie (2009: 258) consider the idea of FM having a single, fixed definition 'highly questionable', with Price (2003: 198) noting it remains merely an 'umbrella discipline due to the nature of its title and the inherent meaning therein'. If there is no firm remit on what FM does, it can in theory perform whatever support role is required within a wide range of circumstances. The presence of large globally recognized FM companies—for example, ISS (International Service System) and Sodexho—would support the notion that possessing fluid definition removes limitations in terms of service delivery. As a service provider serving wider organizations, who themselves are increasingly changeable and 'shape shifting' (Kociatkiewicz and Kostera 2014), this quality of elasticity (Becker 1990) would therefore seem increasingly beneficial. This perspective would mean a transition from the 'emerging' to the 'established' would be much harder to capture, if it could be captured at all.

There is also the view that FM is just a collection of services and functions that are part of general management, and as such, does not represent anything like a recognized discipline. It is not a question therefore of possessing 'identity', but that its activities are dissipated and do not require a specialized

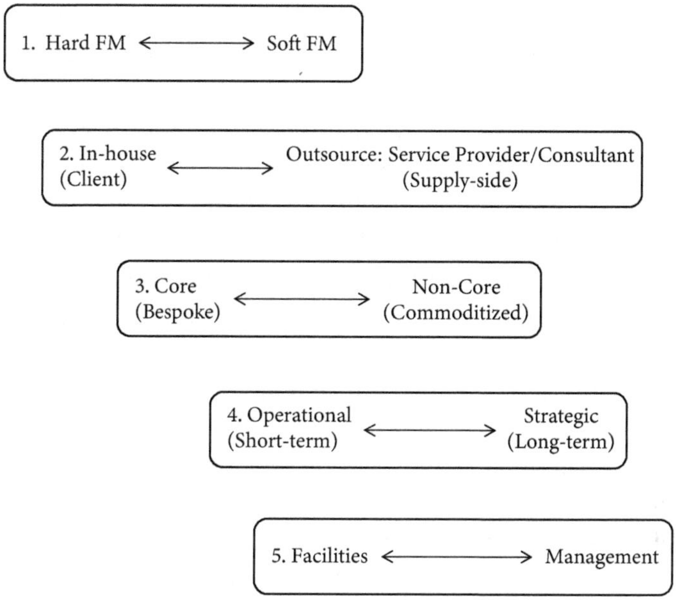

Figure 2.1 The FM definitional spectrum
Source: Authors

approach or knowledge base. It therefore follows that professionalism is not a required attribute (de Bruijn et al. 2001). A more cynical view is presented by Howard (2002), who suggests FM is merely an example of 'management speak', representing organizational hyperbole and reflecting a management fad (McCann et al. 2015); although given a fad apparently lasts for approximately five years (Junghans et al. 2014), FM could well dispute this label after spanning over forty years of activity and debate.

Despite the debate over definition continuing with no obvious sign of resolution, there are, however, some centrally agreed themes associated with FM, although they do serve typically to indicate marked divisions within the discipline—differences often underpinned by business rhetoric. In terms of an analytical spectrum, the main themes in this regard are illustrated in Figure 2.1.

2.1.1 Hard FM versus Soft FM

In this sense, 'hard' FM reflects the *hardware*, the *facilities*: the maintenance and asset management element, which focuses on the buildings and tools rather than the people, supporting the engineering background FM arguably emerged from. Firmly tied into the physical environment, it is associated with large capital investments such as boilers, electrical systems, and connected life-cycle maintenance control. This area of FM is essentially quantitative in nature, with measurement and calculation being major concerns. It is this technical orientation that FM is most associated with, leading it to be also viewed as a reactive rather than strategic function (Ventovuori et al. 2007).

In contrast, 'soft' FM is concerned organizationally with more intangible areas and issues; ones that often resist clear-cut measurement. Examples of soft FM delivery would be contracts for cleaning and catering, areas where there is obvious overlap and indeed different onsite teams often represent them. This latter differentiation is most visible when elements of FM are outsourced—when these divisions appear within the contract to establish what services are being included. Finch (1992: 197) for example implies this division of hard and soft elements of FM in describing how the role of the FM professional is reflected on the one hand in 'the manager who maintains the status quo and concentrates on the physicality of the building' while on the other hand being 'a manager of change, who is concerned with preventing building obsolescence'.

2.1.2 In-house versus Outsource

The delivery mode or mechanism of FM also reflects division, manifesting itself in either an in-house or outsourcing contracting form. Outsourcing is arguably a dominant feature of contemporary FM and as such will be discussed in further detail in Section 2.3. Essentially, an in-house team is part of the organizational complement, and is either a recognized department or has functions associated with FM which are dispersed across various internal departments.

An outsourced arrangement takes the form of an external service provider that is 'contracted in' to the organization to perform particular functions as defined within a contractual agreement. These external service providers are recognized FM companies and vary in size from a single consultant to large, global enterprises. The nature of client-side and supply-side FM should not in theory be operationally different, as both exist to support the wider organization as a service provision. However, this differentiation does reflect variation in terms of the recognition and placement of the function within the organization. For instance, a client-side FM division may be better placed to integrate—or at least communicate more freely—within the corporate structure, as a visible and accepted part of the organization. In contrast, supply-side FM 'has a dual role of delivering designated services to the client whilst also managing the service company's bottom line to the best corporate advantage' (RICS 2013: 9).

2.1.3 Core versus Non-core

One implication of this line of strategic reasoning is of FM being considered as a rather basic or inconsequential organizational function—for 'the language of core and non-core into which FM so easily falls arguably serves to subtly reinforce a perception of unimportance' (Price 2004: 353). Related to the identity issues discussed earlier, the presence of an external FM company may be seen as not only bringing in industry experts, but also as hiring in people to do work that is not key, and therefore can be delivered by 'non-staff'. With the wide range of activities associated with FM, it can be argued that FM holds within its main remit those activities that are non-profit generating and therefore are not crucial to the overall survival of the organization. This is certainly the argument often associated with an outsourcing situation, where non-core elements are identified and contracted out so to allow the company to focus resources on 'core competencies' (Ikediashi et al. 2014). Viewing FM as 'non-core', positions it with being a 'cost-code' as opposed to 'specific service' and can lead it to being viewed as a 'necessary evil' (Price and Akhlaghi 1999). Indeed, in the shadow of recent UK economic recession, where outsourcing was pushed up many organizational agendas as a strategic option, FM runs the risk of becoming increasingly commoditized; a *one-size-fits-all* contract approach. This is despite the range of activities that FM delivers.

However, FM can also manifest itself as a core discipline, depending on the type of organization, its range of operations, and mode of decision-making. For instance, a power plant displaying a high level of technological complexity and risk would arguably consider FM as being a core competence, in comparison perhaps to role of FM in back-office operations, which would reflect far lower levels of organizational impact (Price 2004). Returning to earlier definitions, if FM is focused on 'people, place, and process', then it is hard to distinguish what it does not include in such a wide remit (Thompson 1990); for this would potentially seem to indicate FM having a direct and measurable impact on business, and thus placing it organizationally at a more strategic level. It is interesting to note that within the business literature functions such as project management and IT are similarly considered not so much as standalone disciplines but as functions absorbed organizationally into a similar evolutionary pattern as FM.

2.1.4 Operational versus Strategic

The positioning of FM as either operational or strategic reflects its role as either a reactionary, and therefore short-term service, or a proactive function prospectively. A short-term, piecemeal approach may lead to many organizations viewing FM as 'a solution looking for a problem' (Thomson 1990: 8), whilst a strategic view would place FM as not so much an element which helps the organization survive, but one that actually 'enhances its potential to prosper in a volatile commercial climate' (Tranfield and Akhlaghi 1995: 13).

The operational 'reality' of FM was emphasized in a 2012 RICS report in which managers with responsibility for FM activities were asked about their temporal allocations. This revealed a high percentage of time (52.8 per cent) allocated to day-to-day operations, with only 19.9 per cent being devoted to strategic and planning activities (https://www.rics.org/uk/news-insight/research/research-reports/raising-the-bar-strategic-fm/#III). Another paradox evident within the industry is found within considerations of its predicted future. With little overt agreed identity, it stands to reason that the size of the FM market would be therefore hard to establish (Price 2003). However, it is generally considered as healthy and grew to around £140bn in the United Kingdom alone by 2021 (https://www.cibsejournal.com/news/uks-fm-market-valued-at-120bn/). This is supported for example by initiatives such as building information modelling (BIM)—the generation and management of digital representations of physical and functional characteristics of buildings—which suggest FM taking a step closer to the 'fabled strategic top table' (Heath 2016) of business functions. Nevertheless, the debate concerning FM's future prospects concentrates mainly within the remit of operational improvements as opposed to demonstrating strategic objectives (Atkin and Bilsten 2017).

We also note that the IWFM discussed the 'Death of FM' at the June 2014 Workplace Summit (BIFM FM World, 2014). It could be argued that the confusion arising from a diluted organizational identity makes it difficult for FM to move away from an association with diffused operational service streams. This is emphasized perhaps by the visual impression of 'FMers' (facilities managers) not being 'suited and booted', but often tending to wear fleeces, associating the profession with more janitorial-like functions.

2.1.5 Facilities versus Management

Moving away from hard/soft operational divisions, a new white-collar focus has evolved in FM, essentially concentrating on issues of management rather than facilities. This spans areas such as innovation (Mitchell-Ketzes 2003), agility (Hood 2003), employee retention (Earle 2003), knowledge support (Greene and Myerson 2011), and future requirements (Shrive 2023). Education is currently a large development area. There are only a handful of UK universities offering specialist FM courses at postgraduate level, with most programme pathways being within built environment faculties and departments rather than business schools. This would seemingly be indicative of the weighting of pedagogical importance being more on the facilities than management element of FM. Moreover, it remains that FM is often not viewed as the career of choice for many students but rather an accidental professional pathway; one where knowledge is acquired as much through practical experience as planned education (Hurtado et al. 2014).

It could be argued, therefore, that expansion from a practice that was not firmly established within one recognizable delivery stream has resulted in an unclear identity both within the FM industry and in relation to the primary industries that FM supports. As a service provider, it has been necessary for the FM industry to meet the needs of the wider organizations it services, and thus the drive has been 'mainly about creating a discipline for many diverse services whilst attempting to demonstrate logic in that and the value it can deliver' (Emanuel 2014).

The lack of firm identity—and potentially credibility (McCarroll 2017)—is reflected in the generic name of 'facilities management', which arguably does not reflect a strong professional character. This is illustrated in the IWFM itself, which was rebranded in May 2018 from the traditional British Institute of Facilities Management (BIFM) to emphasize the workplace element and provide wider functional clarification. Other titles such as 'infrastructure management' and 'real estate asset management' have been proposed over the decades (Price 2001, Nenonen et al. 2023). Paradoxically, an American journal declared that in continental Europe, 'there is a trend among service companies to add the two letters FM in order to make them look professional' (Nor et al. 2014: 4), which is possibly why larger FM firms in the United Kingdom have often dropped 'FM', in favour of 'multi-services', to differentiate themselves as experts.

It is apparent, therefore, that many competing functions are evident within FM, and each could claim the title of FM as their own. Holistically, this ability to hold 'one, or more, competing ideas in mind simultaneously and to be able to work with them without either collapsing or condensing them into a singularity' (Cairns 2014: 98) would arguably illustrate the flexibility of the industry, and suggest complexities in creating a universally accepted and recognized discipline. It is therefore useful to examine pre-existing criteria that have been developed to distinguish professional characteristics. Table 2.1 illustrates such a framework applied to FM—to encapsulate not only the current but also planned attributes of the practice. It should be noted that the term 'discipline' applies here to an academic definition, although the term is used across the FM industry in a more generic way to illustrate that it is a practice and profession in its own right.

Table 2.1 Attributes of professionalism

Attribute	Current FM practice	Aspirational FM practice
Intellectual bias: an intellectual discipline requiring good educational background and tested by examination	University and institutional professional qualifications available from Level 4 to doctoral level. Specialized academic journals	Qualifications targeted at school leavers, as a planned career of choice (Hightower and Highsmith 2013)
Private practice: essential expertise, standards of the profession derive from meeting the needs of clients/organizations	Technical language is apparent, but generally in the form of terms associated with outsourcing and delivery mechanism (e.g. 'VFM/in-housing/right-sizing)	Standardized contracting practice (Alexander 1992). Recognized and planned profession of choice
Advisory function: aligned with executive function, following professional prescriptions, or doing ancillary work	FM appears at different levels within different organizations and sectors, from manifesting as part of a wider operational team to representation at board level	Acknowledged need for a common language (Kaya et al. 2004) and legitimate recognized identity.
Tradition of service: based on an outlook which is essentially objective, impartial, and neutral	Recognized British standards on facility management	'Without the knowledge and skills specific to FM, the ability of FM to stand as a distinct profession is diluted' (Tay and Ooi 2001: 361)
Representative institute: Represent members of the profession, safeguard and develop expertise and professional standards	Global presence represented through organized institutional bodies	Recognition of the institute outside of its members
Code of conduct: laid down and enforced by the professional institute	FM code of conduct available via IWFM	New forms of partnership (Alexander 1992)

Source: adapted from: Bennion (1969), as cited in Tay and Ooi (2001: 360).

It could also be speculated that the issues facing FM as a secondary service can be applied to both IT and HR, at least in part (Varcoe and Hinks 2014a), in terms of being subject to the pressures of staying 'lean' and yet competitive (Hui and Tsang 2004). The evolutionary journey of IT 'from back room to board room' spans roughly an equal timeframe to that of FM, approximately initiating in the mid-1970s. As argued by RICS research, 'IT leaders earned their place at the table by developing high-quality enterprise systems, by innovating beyond their competitors, by understanding how to use IT to create strategic advantage, and by ultimately making a difference in the business's bottom-line performance and competitive positioning' (2013: 17). It is this need for the profession to operate strategically that the FM literature commonly reflects; although FM practice often does not, leading to an argument that—given the importance of creativity and holistic knowledge—IT flourished in this regard whereas FM remained static. Such an argument was made at the turn of the century, with Puddy et al. (2001: 505), for example, warning that FM needs to concentrate on 'knowledge' otherwise it may miss that 'really important aspect, which is the ability to use it', with the discipline thus requiring 'freedom of action to take that knowledge and convert it into a saleable product, commodity or service'.

An important element not overtly addressed within Table 2.1 is the importance of a professional 'community'. Community illustrates identity in terms of how people position themselves in a shared manner. In an occupational setting, a profession emerges from the people practicing it identifying themselves as such, creating a sense of belonging (Junghans et al. 2014), and therefore a storied space (Brewer and Dourish 2008). In her investigation on the discipline of humanities computing, Terras (2006) explored the presence of identifiable idols and heroes, mythology, and unique artefacts that contributed to the labelling of a recognizable discipline. It is this classificatory element that dominates the present research, in the understanding that 'professions . . . are no more than systems of belief and practice that have managed to capture a sufficiently permanent niche in a social ecosystem, equivalent to adherence to particular paradigms in science' (Terras 2006: 235).

Within FM, a theory of communities is developing, but it is addressed in terms of the field creating and sustaining work groupings as part of a support function—a notion operating arguably in detriment to the sense of a profession, notably one with a sense of recognizable legitimacy. Through the medium of fairytales, this research aims to uncover the existence of shared stories, following the theory that an individual must 'be able to reference standard traditional stories to prove his or her membership in a particular community' (Zipes 2005: 5). This approach expands past facts not only to explore individual archetypes, but also to help chart how this practice is seeking to legitimize itself as a recognized and authentic discipline.

In summary, it has been demonstrated that FM's confusion over what it is, and what it should be, is one of the only constant themes in a very paradoxical industry. Although it is increasingly positioning itself as a strategic discipline, most FM practitioners are operational in orientation (RICS 2013). As such, much paradox can be identified, for 'FM professes to want to be at the heart of organizational development when clearly many of FM's services are delivered either by external consultants or in-house teams set up as internal consultants' (Ventovuori et al. 2007: 228). However, despite the debate over its importance and function, FM has not withered away but continues to grow as an industry, seemingly being able to mutate and spread to fit different business contexts (Price 2012).

2.2 The Kingdom of FM

Despite the ongoing debate over the definition of FM, it is the connection to the physical environment that the discipline is most associated with, with 'four out of five senior executives agreeing . . . [with the] focus on the maintenance and cleaning of a building' (Thomson 1990: 9). The physical buildings that organizations occupy represent the most visible, tangible, and predictable aspect of FM. Association with the maintenance and improvement of the physical assets of an organization is generally the preferred, if ostensibly still evolving, view of what FM contributes (Drion et al. 2012).

Although there is an increase in flexible working in organizations—that is, work that is time flexible, place flexible, and location variable (Cascio and Montealegre 2016, Hassard and Morris 2020, Manyika 2017)—leading to less time spent within corporate premises, the office itself as a concept is still omnipresent in many people's lives. This represents the place where 'white-collar workers spend much of their time and find at least part of their identity' (Marmot and Eley 2000: 3, also Baldry and Hallier 2010, Pearce et al. 2023). Corporate buildings are a constant, physical symbol of the body that occupies it—a branded space with distinctive symbolic logos and defined modes of association. They are established cyphers that are universal across the globe in their presence: bureaucratic (and post-bureaucratic) machines designed to tell the corporate story (Barnes and Newton 2019, Rippin 2011). As such, they represent 'symbols of power and status that may have deep historical roots in places and fortifications', being equally 'forms of historical architecture' that confirm the role of 'common spaces, the agora, great hall, market forum where members of a society who have sufficient power could mingle and interact' (Price and Fortune 2008: 4).

As illustrated in Figure 2.1, the operational and facilities element of the practice would focus on maintaining and managing the soft and hard elements of the building throughout its lifespan. However, with the earlier analysis emphasizing the balance of people alongside process and place (Tranfield and Akhlaghi 1995), there is an increasing demand on FM to 'deliver by creating different and powerful conversations' (MacDonald et al. 2009: 23). This essentially expands FM's focus on the bricks and mortar to include, not just the space, but also the business activities within the space, focusing on the *management* and not just the *facilities* element of FM. Despite, the FM industry within the United Kingdom arguably placing a time-honoured 'emphasis on measuring cost per unit areas of space rather than seeking evidence of effectiveness and outputs' (Price and Fortune 2008: 2), increasingly this argument is expanded to take account of not only spatial functions but also the nature of the organizational inhabitants (Strati 1998), with FM often creating storylines for the range of corporate actors it serves.

In comparison to physical environment, the high impact of social factors in terms of employee productivity and satisfaction was highlighted in the famous Hawthorne Studies (Pennock 1930), although the controversy over reliability issues that followed these experiments should also be noted (Cairns 2003). Environmental psychology is another important issue when considering such social and material relations (Mehrabian and Russell 1974), alongside a role for concepts from human geography and the sociology of space (Hirst 2011), with the temporarily of workplace arrangements also being the subject of much debate in this area (Hansen et al. 1976, Hancock and Spicer 2011). Indeed, the relevance of these various issues finds much resonance in the title of Mazumdar's (1992) paper 'Sir, please do not take away my cubicle: the phenomenon of environmental deprivation'. This title not only echoes the general power dynamics that can inhere in office layouts, but also those particular to the rise of open-plan spaces (and the decline of cubicle working) at that time; an argument which supports Lefebvre's (1991) concept of space and the suggestion it is 'shaped by capitalism' and is both 'a social product' and 'productive of social relations' (Hirst 2011: 767).

Maintaining, but also controlling and manipulating, the physical space that people occupy links to the practice of FM creating and maintaining the realities of others. As Küpers (2011: 45) argues, 'with all their artefacts, ongoing practices and relationships, as well as with ambiguous experiences and meanings, places of work have multifarious impacts on those dwelling in them'. In this manner, FM becomes the physical embodiment of the organization itself, locating the profession uniquely within organizational dynamics. The association of the built environment with FM reflects an active, social relationship as opposed to a belief that 'buildings are populated by mere Letterset zombies, some satisfied, some miserable, but all propelled through the built environment, involuntarily pushed this way and that by somebody else's design decisions' (Shove 1991: 11).

In recognition that a building is not just a static entity but also a process of perpetually evolving narratives (Alexander and Price 2012), FM would then play the role of the corporate storyteller, translating the organizational mission, vision, and therefore tale from strategic to operational levels, making the story into a physical reality to be interpreted by the individual inhabitants of the occupational landscape. The audience spans the organizational hierarchy, and the stories are often projected out to the external realm 'proclaiming a narrative of identity, designed to legitimize past, present and future decisions and strategies' (Rippin 2011: 720). In support of Sims (2003: 14) reference to middle managers being aggravated by their lack of organizational legitimacy, FM managers are similarly 'expected to weave a narrative for those who work under them to believe in and be able to give a coherent narrative account of their performance to their superiors'.

Additionally, many organizations are introducing 'spatial rhetoric' (van Marrewijk and Yanow 2010) to define their modern internal topography, deploying novel concepts such as corporate neighbourhoods and villagescapes instead of traditional bounded notions of department, section, or unit. This is in an effort to foster interdisciplinary relationships and mitigate office fiefdoms in an organizational world habitually associated with allocated space size and related symbolic power. The concept of the neighbourhood aims to foster relations that supports the collective commitment and loyalty to the organization, creating communities at work (Larsen et al. 2011) and supporting social and individual motivation (Dale and Burrell 2010).

The increasing realization of the effects of space within organizational dynamics is interesting for analysing FM. For instance, highlighting the idea that organizations are rarely 'holistic' in their operations—but instead often take recourse to discrete silos in structuring actions—suggests that any concept of FM as holistic is consequentially lost. If the concept that 'organizations are essentially complex systems characterized by the interdependence of the social and the physical' (Price 2012: 3) fails to be recognized, why would you need a discipline to not only support and maintain the business it but also to foster its growth? This may illuminate why FM has never attained an accepted identity in the United Kingdom but has nonetheless thrived as an industry (Jenson et al. 2012).

Another view would be that, in this setting, FM—cast as the metaphoric 'person in the fleece'—could assume an important role in representing a comforting belief that the organizational world is a predictable and reliable place (Gabriel 2013). This role is accentuated in an era where 'jobs, formally seen as being for life, are more often than not merely temporary and many disappear virtually without notice, together with the factories or office or bank branches that offered them' (Bauman 1995: 265). It may be that this associated role has sustained FM within the organizational community, becoming the underlying symbol of things that can be controlled in a fluctuating environment. For example, it is 'hard' FM—in the guise of maintenance—that is visibility present when the physical environment needs repairing. This arguably symbolizes an identifiable problem that can be remedied when the chaos that organizations fight so hard to control becomes manifest (Gabriel 1995). Alternatively, it could also point to FM serving to highlight the 'rejected aspects of what the rational organization would like to make itself to be like' (Kociatkiewicz and Kostera 2014: 12).

A further view still is that modern working practices reflect an increased focus on knowledge co-creation and network formation. The term 'knowledge worker' was first applied to professions such as doctors, lawyers, scientists, and academics (Drucker 1999), but has since evolved to include a wider remit of the 'creative class' (Grantham 2003). Finch (1992) first highlighted the rise of the knowledge worker within the FM literature, in relation to analysing the property crisis of the 1970s and 1980s. This was against the backdrop of an industry-based economy being progressively replaced by a more knowledge-based one (Greene and Myerson 2011).

Additionally, Leaman (1992: 19) suggested the creation of a disciplinary knowledge base 'will ultimately, perhaps in ten years' time, define . . . FM and provide the foundation for professional legitimacy'. This point emerged again nearly a decade later, when Nutt (2000) highlighted the importance of knowledge management in FM evolving alongside financial, HR, and environment foci. This would indicate that Leaman's prophecy endured as such, with the call remaining for FM to identify, capture, and promote knowledge (Pathirage et al. 2008), but also to do so to support a more self-motivated and self-managed workforce. As a support service, this would be vital, not only for what FM would need to deliver, but also for what the profession would become.

Transferring this concept into a wider setting, Moravec uses the term *knowmads*, who are 'context workers, applying what they know into new contexts to create value within different organizational and social configurations' (2013: 81). These workers are not place-bound but adapted to work in line with the boundaries of work, family, and leisure that global, economic and technological advances have blurred (Steiner 2006). Further, they are not traditional nomads in terms of going where the work/capital is—to later return home—or the *homeless minds* of those who have their roots 'within their suitcase' (Czarniawska 2014), but rather those who have no recognized sense of belonging either to a physical space or symbolic 'home' through company allegiance or logos. The influence of these new recruits, who are arguably selected for profit-based orientation (Dale and Burrell 2010) and would seem to defy any concept of a psychological contract, thereby effecting the possible creation of a community. This element of perceived separation can also lead to a sense of estrangement from workplaces (Hodder 2020) and have negative impacts in areas such as employee performance and co-working (Van Der Lippe and Lippěnyi 2019).

These elements are the hallmarks of the increasingly turbulent work environments FM professionals are charged to create and support, with FM practice itself becoming the prototype of its creation. FM is therefore intrinsically tied to organization and how it emerges 'in the symbolic co-construction of place-meaning which is bound up with the experience of space' (Lucas 2014: 5). The presence and increase of knowledge workers both within and as clients of FM practice is recognized, although the operational dominance of FM indicates that how to recognize, capture, and promote this aspect is lacking (Pathirage et al. 2008).

It follows that the demand on space is therefore changing in the wake of the growing trend to work anywhere, anytime, in line with work–life flexibility (Ross et al. 2016). With the onset of the fourth industrial revolution (Lim 2021), FM is increasingly affected by what is termed the *third place*. Home is considered the first place and work the second, with the traditional theory that third places are 'other' spaces (Philips and Rippen 2010). They increasingly represent the anchors of community life (Oldenburg 1989) where people are encouraged to gather. In terms of workplace analysis, third places are recognized as 'life between buildings' (Gehl 1980 as quoted in Larsen et al. 2011: 85), a 'kind of nowhere and everywhere at the same time' (interview response in Varcoe and Hinks 2014b: 14). Third places are illustrative of the impact of employee choice, but also the importance of the chance meeting, spontaneous conversations, and linking back to the power of networking (Puddy et al. 2001). The idea of work no longer being place-bound but completed with priorities of convenience and time management are also factors in line with a nomadic workforce whose members take their work into a range of public settings and community spaces, transcending the traditional role of the lobby, lounge, or waiting room (Breure and Van Meel 2003).

As early as 2003, on surveying two large American companies, Grantham found that 20 per cent of employees carried out tasks in third places for an average of ten hours or more per working week (Grantham 2003). As such, it follows that the organizational neighbourhood as analysed earlier has correspondingly evolved past the physical workplace. The Covid-19 pandemic has led to a reconsideration of job design (Kramer and Kramer 2020), both at collective-organizational and individual-worker levels in many industries. This is illustrated by the variety of flexible working arrangements such as job sharing, staggered hours, flexitime, and hybrid working (https://www.cipd.co.uk/knowledge/fundamentals/relations/flexible-working/managing-flexible-working-requests-guide).

Symbolically, the transitional areas where people, by definition, gather (Oldenburg 1989) act as a place where workers can 'reshape or re-order their sense of self' (Lucas 2014: 17). In this sense, they can be affiliated to the concept of the *unmanaged organization*, which Gabriel (1995) describes as a 'dreamscape' that is present in all organizations. Essentially, this lies outside the remit of managerial control and is where desires, emotions, and fantasies manifest. Rejecting traditional views of fantasy in favour of a 'third way', in Gabriel's work this flips and reframes dominant managerial practices so that emotion and *uncontrol* override the drive to the otherwise supreme rationality.

How FM will support these movements, if indeed it can support them, has arguably been an issue consumed by the wider impact of technology (Marmot and Eley 2000). The idea that IT has the potential to render physical workplaces obsolete (Delbridge and Sallaz 2015) lingers, and is emphasized in the era of Covid-19, where technology has not only enabled flexible working, but also an attempt to replicate organizational culture and a sense of belonging online, through apps such as Facebook (Hodder 2020). However, this can be positioned as an example of the power of knowledge, information, and relationship management that is now being increasingly demanded in organizational life. This emphasizes that FM is increasingly a practice that relies on fragmented relationships and organizational discontinuity as it supports the cast of 'imagined communities' (Dale and Burrell 2010: 37) of corporate life, which itself is a manifestation of the wider *liquid* quality of modernity (Bauman 2007).

The idea of the physical community is thus no longer centred so much on a physical space in an integrated and tightly controlled environment. The 'real' space is transferred by daily routines to spatial practice (Wasserman and Frenkel 2015). Borrowing from Lefebvre's theory of space (1999), FM reflects a concentration on conceived space—that which is designed and built. This contrasts with lived and perceived space, which addresses social interactions, emotions, and symbolism—where 'the imagination seeks to change as appropriate' (as quoted in Petani and Mengis 2016: 73). This is closer to the notion of *emplacement*, a community bond and sense of belonging that is social and set essentially beyond physical boundaries (Dale and Burrell 2010).

The changing demands of physical space and increasing volume of knowledge work, alongside the longstanding divisions, would suggest FM is in a state of flux—a state possibly well accounted for by the concept of 'liminality'. Developed by French anthropologist and folklorist Arnold Van Gennep in 1909, the concept was later built on by anthropologist Victor Turner (1920–1983) to denote rituals where an individual enters a state of 'betwixt and between' (Johnsen and Sørensen 2015: 321) as they move social positions (such as in initiation rites). Specifically, it implies a quality of ambiguity or disorientation that occurs in the middle stage of a rite of passage, when participants no longer hold their pre-ritual status but have not yet begun the transition to the status they will hold when the rite is complete. During a rite's liminal stage, therefore, participants are positioned at the threshold between their previous way of structuring identity or community, and a new way, which completing the rite establishes. In Figure 2.2 we show this diagrammatically in a model adapted from Czarniawska and Mazza's (2003) work on consulting as a liminal space. Here three distinct phases are apparent: separation from the normal triggering entrance into a liminal realm before emerging as something new that is adapted to comply with new social norms (a child to adult in terms of initiation).

Applying this to the physical environment that FM is charged to create and maintain, liminal spaces can manifest as domains which are 'not inscribed with specific activities, rules of conduct or dress codes' as rather they are 'fill-in or add-on spaces that fall outside of the geographic grid of identifiable

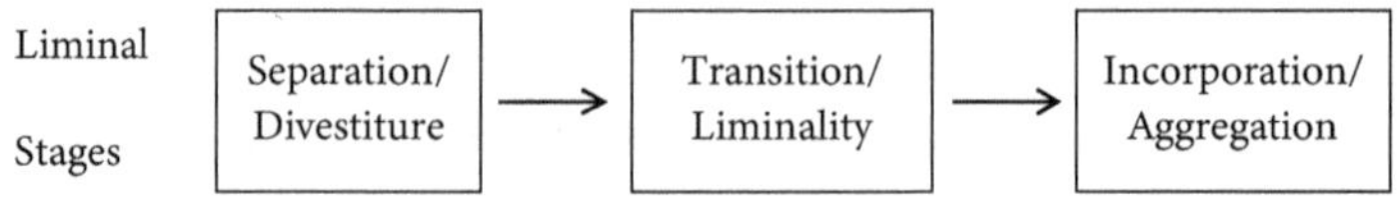

Figure 2.2 Stages of liminality

Source: adapted from Czarniawska and Mazza (2003).

social and organizational events, conducts and functions' (Iedema et al. 2010: 1). As such, they represent neutral spheres such as anterooms, hallways, and lobbies: third spaces where people pass through but do not formally live or work. They are places of transition, the *other*—areas hidden in plain sight (Kociatkiewicz and Kostera 2011, Lam et al. 2018) and whose opacity is reflected in and through the kinds of social interaction that occur within them. Their presence is permanent, but their experience is transient and temporal. It is these domains that are increasingly being purposefully built into office space in particular—to encourage informal knowledge transfer, networking, creativity, and a sense of belonging (Lam et al. 2018).

In organizational studies, the concept has been utilized metaphorically to represent, for example, areas between established fields of inquiry (Hassard and Cox 2013) or to analyse identity reconstruction (Beech 2011). In this view, the liminar is 'socially if not physically invisible' (Beech 2011: 287), and 'temporarily undefined' (Czarniawska and Mazza 2003: 271) which reflects the earlier analysis of FM as being hidden (Becker 1990) and marginalized (Barrett 2000). Other elements, such as being powerless and displaying complete obedience to corporate elders (Jensen et al. 2012), indicate a more cynical view of contractual FM. A further element of the liminal space is that of reflection, 'between history and imagination' (Goulding et al. 2018: 899), that permits consideration of previous social roles to allow a new identity to emerge and activate the third stage of integration. This new identity supplies meaning not only for individuals but also for the community into which they are now integrated.

In essence, liminal spaces are dangerous and yet creative, unsettling and yet sites of togetherness (Lucas 2014), displaying paradoxical traits as identified within FM (Figure 2.3). To move past the liminal stage, traits which are valued and recognized by a community need to be established—being reconstructed through 'symbolic action and interaction within social contexts' (Beech 2011: 290). In short, the liminal space reflects the FM storied space, where an agreed identity of the profession is so fluid it is unstable, creating many diverse and disjointed stories that have little collective consensus. The creation of an accepted narrative of one storied space—that dictates the accepted behaviour of this collective—would then allow the discipline to become more accepted and recognized in the wider organizational plane and enter the final stage of liminality: incorporation.

FM in its role as a secondary service is arguably salient to this, for: 'every organization has its unique personality, and the facility department is a reflection of that personality. All facility managers share some common characteristics in management style; however, the emphasis varies with the type

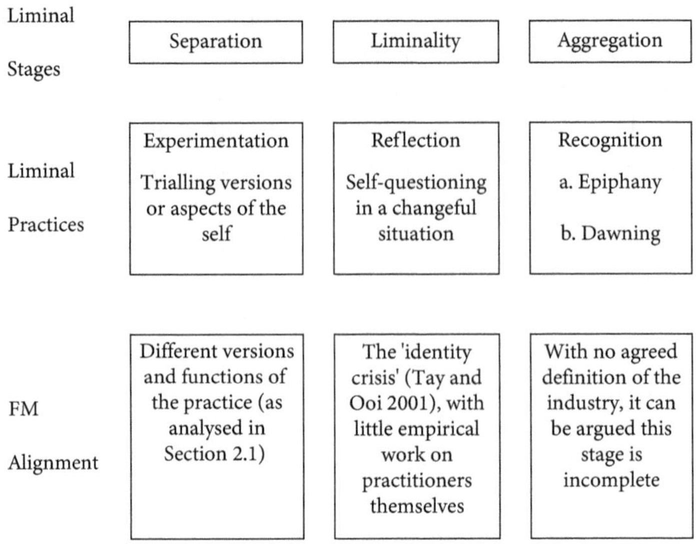

Figure 2.3 Liminal stages aligned to FM
Source: adapted from Beech (2011: 290).

of organization' (Cotts et al. 2009: 21). It can be argued therefore that an eternal presence within the liminal phase is not necessarily detrimental, but allows a fluidity that is established within the unanchored identity described in Section 2.1. Being *other*, by being separated from a community can be liberating, with sufficient people located outside a community leading potentially to the creation of another: 'A special sense of community with the others in the limbo that has little to do with identity—rather a shared sense of alterity' (Czarniawska and Mazza 2003: 273); that is, with no sense of urgency to be reabsorbed into the mass. Thus, it can be argued that by deploying the concept of liminality we can facilitate the investigation of patterns of storied spaces of FM in novel and creative ways.

2.3 The Quest for Meaning

Within this chapter, debates within the FM industry have been highlighted, and the working environment has been identified as the physical and metamorphic palace of FM. A theme of liminality has been introduced in terms of analysing the different storied spaces of the discipline. This final section discusses further important facets of FM—outsourcing and operational forms within the service delivery mechanism.

Outsourcing is essentially the 'contracting-out of services that were previously performed in-house to an external service provider for a fee as a means of increasing organizational efficiency and effectiveness' (Ikediashi et al. 2014: 472). As such, it is mainly associated with non-core tasks, and metaphorically 'removing' them to allow the organization to focus on 'core competencies'. It is this reasoning that arguably characterizes FM as an easily replicable service as opposed to a strategic discipline, as discussed in Section 2.1. With outsourcing being promoted as a strategy on many organizational agendas, this has served to emphasize FM's potential as an option in this respect.

Correspondingly, this quality increasingly dominates the image of FM as a practice, with the range of services it offers under forms of outsourcing agreement being seemingly uncapped. This arguably dilutes the 'function' of FM even further—positioning the discipline with a range of low-skilled services that can be figuratively boxed up quickly and sold off cheaply as a collection of non-core activities: 'What they should really say is *Veneer FM*, because they provide a thin layer of co-coordinating management over the top of these silo'd series, giving a single point of contact and an attempt at reducing the loss of values between the component parts' (Varcoe and Hinks 2014a: 9).

Long-term strategies are sacrificed for short-term gains within outsourcing contracts that are typically of the 'three–two–plus model' form (i.e. a three-year contract with an option to extend by a further two years if agreed)—which can stifle the creation of 'intelligent cooperative' forms of agreement that are necessary to support 'service delivery ecosystems' (Varcoe and Hinks 2014a). This type of operation, which is by no means universal,[1] adds weight to the assertion that 'the FM industry knows the cost of everything, but the business value of very little' (RICS 2013: 6). Therefore, the connection of the FM industry to outsourcing is one that is a source of controversy: 'The perception that FM equals outsourcing is just that—a perception—and a highly distorted one at that . . . take every opportunity to refute the notion' (Drion 2012: 258).

Not only does this arguably intensify all the existing divisions as illustrated in Figure 2.1, but it also accentuates three distinct roles, and therefore further storied spaces, within FM: that of the in-house manager (who is part of the organization) and two external entities (of private FM service providers and individual FM consultants). In doing so, there is a delineation of the

[1] For example, alternative arguments—regarding innovation potential and the 'safety of the environment'—are made in the comparative study of outsourced and in-house FM by Mudrak et al. (2005).

latter two types of FM as they fall outside the generally considered fraternity; those who do not 'belong' (Dale and Burrell 2010). As outsource providers completing tasks within a client organization, such actors by default have not been through the selection and recruitment process of the host organization, and as such they remain 'other', an external entity. They are thus separated, illustrating the first phase of liminality (Figure 2.2). In this manner, FM operatives could be said to parallel consultants, as analysed by Czarniawska and Mazza (2003), in that they are structurally invisible and exhibit the stages of liminality as identified in Figure 2.3.

This segregation, and associated confusion over identity, is exaggerated when FM provider teams are based onsite and identify more with the client organization than their own company, a practice highlighted in the industry rite of 'the changing of the fleece' (Cowls 2015: 24). This also relates to practices under TUPE,[2] displayed most acutely when a service provider loses the contract, and its staff can be transferred to the incoming supplier and with a new agreement. This can see such staff accordingly 'rebranded'—literally changing the 'fleece' synonymous with operational FM (Cowls 2015). Therefore, there is a 'separation' of FM staff, not only in terms of physical location from the parent company, but also in terms of identity, as they wear the host organization's branded clothing when working on the client's operational site. However, the client firm does not recognize such FM work teams and streams as part of their organization, for they are merely a contracted and temporal element of the operation.

This could therefore be illustrative of the implicit quality of liminality within FM—being arguably a negative one in terms of identity and belonging, but also allowing a certain freedom to construct new identities as the contract allows. In the context of this research, the quality of liminality essentially provides a window for new stories to be created and old paradigms replaced. The contracts that follow such a change are positioned as story 'scripts'—forecasts for the future that cast people in allocated roles to play. The knowledge that is contained within such agreements is not so much that metaphorically signed on paper, but rather which is embedded in the experience of personnel and created through social interaction (Puddy et al. 2001)—a form of knowledge that creates new negotiated storied spaces. Taking into account the transcendental and fluctuating nature of FM, seen in its

[2] TUPE refers to the 'Transfer of Undertakings (Protection of Employment) Regulations 2006', as amended by the 'Collective Redundancies and Transfer of Undertakings (Protection of Employment) Regulations 2014'. TUPE rules apply to organizations of all sizes and protect employees' rights when the organization or service they work for transfers their labour to a new employer (http://www.acas.org.uk/TUPE).

unanchored definitions and liminal processes, these storied spaces and their associated events would indicate that 'crooked tales, twisted sentences, complex plots [that] still exist, hidden in the margins and in everyday experience' (Kociatkiewicz and Kostera 2014: 2).

This approach firmly recognizes that FM is a service, and as such carries distinguishing characteristics of interaction, heterogeneity, and perishability (Clark and Salaman 1996). The interactions implied by the role of FM as a service provider also led to an element of intangibility, which is where this research is positioned. As we explain, FM as a practice is arguably pulled between the 'enhancements' of contractual management; which are associated with the operational, technical, and increasingly reactive (Barrett 2000), and the 'enchantments' of corporate desire; fulfilling client wishes, for example, through realizing the mystical state of 'Total FM' (Atkins and Brooks 2021), by controlling the 'army of fairy folk' (O'Briain 2007), or in crafting tales of impossible success to create a legitimate 'cultural' role (Crupi 2016).

To clarify these concepts, the accepted definitions of both enhancement and enchantment need to be declared and considered. Enhancement is a popular term within the FM industry, illustrated by a quick keyword search in the main journals (Coenen and Läuppi 2014), with the *Journal of Facilities Management* containing 175 articles associated with the word enhancement and *Facilities* carrying 578. Enhancement carries with it the association of being a new and improved version of an existing item (Law 2009), and as such, is arguably familiar within FM as it continues to change, seeking an accepted and legitimate function in organizational life. Enhancement is reflected in contractual agreements that state, 'no surprises are important for clients' (Jenson et al. 2012: 202), which can be taken to mean both beneficial and detrimental elements, in favour of maintaining the status quo.

Although the practice of FM could be described as a wide collection of often seemingly disparate offerings (Section 2.1), the common denominator of providing a service remains. In this sense, service work links 'labour to the reproduction of social and symbolic as well as material realities, and consequently to the potential for the (re)enchantment of the workplace' (Endrissat et al. 2015: 1556)—a state that is inclusive of both the provider and the client. Given this sentiment, it is therefore somewhat surprising to find the concept of enchantment completely absent from the current FM literature. The concept is, however, gaining traction from areas such as media/art studies (Macleod 2008) and marketing (Woodside et al. 2008), and also making forays into fields such as change management (Boje 2011), where it starts to show some overlap with organizational storytelling. In organization theory, it is considered exceptional conceptually (Ritzer 2005), but has gathered momentum in line with its evolving relationship with disenchantment spanning back to Max Weber's (1922) thesis on *'Disenchantment (*Entzauberung*) of the world'* (Endrissat et al. 2015).

Enchantment is interpreted as 'an aura of authentic presence, resisting rationalisation . . . [it] evokes the personal, the socially meaningful, and even the sublime' (Endrissat et al. 2015: 556), which may help create further understanding of the different storied spaces within FM. Appreciation of enchantment follows analytically from earlier discussion of FM creating and maintaining the 'realities of others'—realities which are increasingly separated from the material built environment to form 'imagined communities' (Dale and Burrell 2010: 25). Taking this position to its extreme, it can be argued for instance that the rational is suspended in illogical contracts and enacted by magical beings who remain invisible, as FM seeks to balance and restore social connections within and across organizations.

Enchantment also carries with it a connotation of malevolence in terms of presenting something that is not real but rather a manipulating force that can fool and dupe. In terms of an organizational context, for example, this could be illustrated in the 'illusion' of worker autonomy (MacDonald and Sirianni 1996), but it perhaps has wider ramifications on the organizational landscape with 'hard, visible, tangible work being replaced by work that relies on the manipulation of images, signs, aesthetics and emotional displays' (Gabriel 2005: 23). Applied to FM, it is most readily illustrated in the rise, and associated criticism, of global FM service providers who sell any service, anywhere, at a price. This is 'veneer FM' as discussed earlier by Varcoe and Hinks (2014a)—which positions FM as the enchanter; seen in examples such as Heath's (2016: n.p.) observation of large FM providers keeping up 'the smoke and mirrors act' because they are 'the only thing keeping the emperor in new clothes'.

Endrissat et al. (2015) offer three strategic elements in identifying characteristics of the otherwise quite ambiguous nature of the enchantment of service work, as detailed in Table 2.2. How this can be mapped into FM is then highlighted to help understand how this concept is already apparent, although not overtly recognized, within the practice. The first two elements lean towards the darker side of enchantment, aligned to disenchantment, with the final 'authentic' elements being affiliated to the second definition of the theory (see Ritzer 1999).

The first strategy, focusing on normative control to ensure consistency, would arguably reflect the view that FM is variously operational, a commodity, and an 'off-the-shelf' facility, rather than a legitimate business discipline.

Table 2.2 Strategies of service work enchantment

Strategy	Characteristics	Focus	FM alignment
Normative control to mitigate service ambiguity	De-skilling or standardization resulting in predictable and disenchanted routines	Industrial viewpoint: Ritzer's (2000) 'McDonaldization'	Commodification of FM and 'changing of the fleece' (Cowls 2015)
Pre-approved fun, or highly controlled enchanting experience	Poor substitute created—elements of 'institutionalization' (Islam and Zyphur 2007)	Fantasy and spectacle (Boje and Baskin 2008), Disneyfication (Boje 2011), Disneyization (Bryman 2004)	'Veneer FM' created (Varcoe and Hinks 2014a)
Ability to generate authentic sense of presence and purpose	Using cultural, interpersonal, and artistic means to create reality and authenticity, whilst avoiding harsh reality of low-paid service work	Foster creativity and participation. Self-expression. Authentic—but necessarily true?	Relationship management; partnership approach; recognized value and dialogue

Source: adapted from Endrissat et al. (2015: 1559).

This links to the earlier analysis of the rise of outsourcing and its corresponding association with FM. The appearance of global FM service providers has arguably intensified this interpretation, which can be unpacked by reference to Ritzer's concept of McDonaldization (1993) and its four qualities of efficiency, calculability, predictability, and control. In terms of FM delivery, this is mirrored at a strategic level in material factors such as increased use of bundled contracts and the evolution of 'Total FM' packages. At tactical and operational levels, the de-skilling of some front-line operatives allows for a 'universal offer' to be made for task completion; with, for instance, the security function, rather than focusing on a single role, also completing cleaning tasks under an agreement to employ multi-purpose site operatives.

Similarly, building on a second strategy of 'pre-approved fun' and creating a highly controlled 'enchanting experience' leads to consideration of the concept of Disneyization; or the process by which 'the principles of the Disney parks are coming to dominate more and more sectors of society' (Bryman 2004: 1). In terms of FM, this reflects practices of transforming settings, situations, and environments into something simplified, controlled, and 'safe'.

Much of Bryman's analysis of Disneyization concerns corporate practices and processes replicating elements of 'dedifferentiation of consumption', 'performative labour', and 'theming'.

Dedifferentiation or hybrid consumption is the general trend whereby the forms of consumption associated with different institutional spheres become interlocked and are thus increasingly difficult to distinguish. This reflects a process by which structures or behaviours that were specialized for a specific function ultimately lose their specialization and become simplified or generalized. For FM, there is an obvious overlap with McDonaldization here, for example, in allowing a new range of services to be bolted on to the original contract.

Performative labour follows the scripted roles that can be located within contractual agreements, where employees are given roles to play akin to a staged theatrical performance. By application to FM service delivery, this can be seen not only in the initial impressions of a building (e.g. a clean and welcoming reception area), but also in the symbolism of human behaviour and interactions (e.g. a smiling, helpful, and approachable receptionist). The staging element is also witnessed in FM materiality and rhetoric; for example, with 'front-of-house' and 'back-of-house' areas, staff decked in matching uniforms/costumes, and work underpinned by the semantics of having a 'role' as much as a job. Explicitly scripted elements are also discernible in telephone responses, voicemail responses, and out-of-hours email notifications.

> We ask for service knowing that the one serving us has in fact had to rehearse a variety of 'lines' and that in being served we are equally practiced in appropriate lines of response. The most profound irony is that in such engagements we all operate under the fictions of spontaneity in conversation but judge the server by the criteria of style in performance—that is, by the appropriateness of the routines performed and by the style by which the service is carried out. (Abrahams 1978: 178)

In addition, the condition of theming also concerns 'the application of a narrative to institutions or locations . . . which provides a veneer of meaning and symbolism' (Bryman 2004: 15). The narrative in this case is the discipline of FM itself. The multidisciplinary nature of FM could arguably dilute theming, which in turn may exacerbate the confusion over FM's identity within contemporary organizations, as discussed in Section 2.1. In terms of the built environment that FM focally manages, theming would align with the creation of a physical landscape that reflects the identity and regimes the company represents (van Marrewijk and Yanow 2010). By applying the concept of theming, associated identities of the industry can be illustrated, and notably by exploring some of the metaphors the FM industry itself deploys. Notably, the idea that FM is the Cinderella industry is widespread (Gaughan 2018)—highlighting associations for example with the 'dirty work' debate (Cassell and Bishop 2014), the 'from the ashes' metaphor (Ciacchella 2015) and the notion of 'hidden' service (Cabrelli 2010).

The highly controlled enchanting experience is therefore one in which the client witnesses a stage show of actors performing pre-planned routines within an artificial environment. Returning to the definition of enchantment, this element appears a manipulating force, with the argument suggesting the client is being duped into believing the 'veneer' (Varcoe and Hinks 2014a)—with FM as a service becoming increasingly disenchanted by creating the delusion. This in essence reflects the Weberian notion of rationalism, where social processes progressively become standardized to achieve specified goals. Ultimately, as presented, this scenario supports the creation of an 'iron cage' (Weber 1947), where each actor—for example, service provider and service receiver—becomes locked into expected roles, paradoxically stifling service innovation as they both 'become constrained by the very structures they have created' (Islam and Zyphur 2007: 773). In such circumstances subversion can appear, such as game-playing within service contracts (Hopland 2015).

The third strategy identified in Table 2.2 is the ability to generate an authentic sense of presence and purpose, which on the surface seems far removed from the above analysis. This study proposes that if 'magic' has been largely replaced by 'science' in Weberian logic, then the remaining characteristics and foci may have been repositioned. Supporting arguments for this position come from Bowles (1989), who link the demise of the social role of the Church to an upsurge in people seeking to receive similar meaning and structure in life through work organization. This approach resists rationalization, and instead involves 'using cultural, interpersonal and artistic means to create the sense of reality and authenticity' (Endrissat et al. 2015: 1559)—a position which draws upon more intangible elements of the lifeworld.

In a five-year international project into widespread changes in organization, production and work practices, Casey (2004) highlighted this element of enchantment, specifically through exploring spirituality at work, with a participant stating, for example, 'we need to do something about the soul being marginalized in organizations. We need a wake-up call . . . when we allow the soul to be expressed and nurtured in the workplace, we learn to understand the world differently' (2004: 68). This can create what Collins and Porras (1994) refer to as a 'cult-like' consciousness of intense loyalty and pride, which follows through to the employee–customer relationship as a source of mutual satisfaction (see also Solnet and Kandampully 2008). Rather than reflecting isolated and protected containers of rationality (Jenkins 2012), organizations themselves contain rituals, symbols, legends, and traditions (Bowles 1989), and are affected culturally and relationally by the various 'actors' that make up their collective 'network' (Law and Hassard 1999). In Casey's (2004) study, other elements connected to organizational enchantment were a growing interest in 'magic', 'expressivism', and 'counter rationalities', alongside a wide range of 'alternative' practices that would be traditionally excluded from management theory and analysis.

If the shedding of the physical environment is combined with the theory that work (and the workplace) is becoming transcendental in terms of meaning and perception, this approach could be proposed as one of the vehicles to solidify the FM discipline and influence the next evolution of its service contracts. This approach places FM in a position that resists commoditized/off-the-shelf delivery through scripted staff, and emphasizes instead a service of creativity, and knowledge of the business at a wider, more strategic, level. It supports the conceivable legitimization of a new organizational practice that operates successfully within an evolving network-orientated and boundary-less organizational landscape. This offers a rethinking of FM and its storied space—one proposing insight into how FM practitioners are reframing the roles they cast themselves into within the liminal spaces of the discipline. It supports a call for the FM manager to 'find the means to engage with the different socially constructed realities of different groups of actors' (Cairns 2014: 103) and in so doing seek to bring back the 'magic' amidst a 're-enchantment impulse' (Casey 2004).

Similarly, Endrissat et al. (2015: 1559) refer to this as 'enchanting work'—or to at once 'mystify and be real, to promote the meaningful work relations that foster creativity and participation, and put customers, workers and wider community members in less formalized relationships with each other'. Organizational demands and constraints such as deadlines, targets, and goals are balanced by the ability to shape and affect corporate surroundings; as opposed merely to responding to mechanical elements of normative control (Boje and Baskin 2011). What is referred to as enchantment in the context of this research is seen not to be at odds with the dominant forms of bureaucracy that FM operates within, but as a growing element within the evolution of modern work practices. As Casey (2004) points out, it can be accommodated in organizational life if it is perceived to contribute to the cornerstone principles of bureaucracy—such as effectiveness, productivity, and stability—albeit under the attitude of accepted risk. This reflects the underlying principle that 'the fixation on the tangible is another spell that must be broken' (Tyler 2011: 155).

Enchantment is therefore allied to an element of wonder, and its possibilities explored through social connections and the togetherness of a community. The concept of liminality, as discussed, would indicate that this element is already present within FM practice, but has not yet been fully explored. It may be representative of how enchantment has not entirely left the world to make way for the scientific (Weber 1930), but rather has changed form, and is visible notably in organizational stories (Bennett 2001). These are representative of how 'people are participants in an ongoing process of creating meaning, drawing elements from an overabundant reality, constantly in flux, and reading the world they create as result of the relationships and apparatus/discourses of those practices in which they are seamlessly embedded' (Boje and Baskin 2011: 415). In this context, the narrative presented in organizational tools such as contracts are arguably narratives of disenchantment, with individuals who are tasked with practical implementation being allowed to act with personal discretion and creativity, that is, to surpass and exceed written terms and conditions. Enchantment, like liminality, is arguably best appreciated in a temporal setting.

The application of enchantment expands the storied space of FM into relationship management. While Kaya et al. (2004) identified three essential FM roles in this regard—a translator sitting at a strategic level, a demonstrator to empower the end-users, and a processor to maintain and support changing business needs—this research explores the role of a fourth, a storyteller. It specifically focuses on the internal organizational story from the perspective of FM practitioners, and how their storied space is understood and manifests in terms of reoccurring patterns, plots, and archetypes. This is where stories are formed that portray not just the mechanics of delivery, but the beliefs and norms which are repeatedly endorsed to become storytelling folklore (Solnet and Kandampully 2008).

2.4 Coda

This chapter has introduced and analysed debates within FM over its identity and therefore function. As a secondary service, it forms its identity though special interactions, 'by displaying one's identity and gaining verification from others . . . or through positioning oneself in relation to others' (Rostron 2014: 96). This is visible in many other secondary services, such as HR and IT, although identity issues are particularly discernible in FM, making it a very salient sector to analyse in this regard (McCarroll 2017).

The concept of liminality is identified as an important definitional aspect of FM—one that can add further understanding to the storied spaces of the discipline and its practice. As traditional, transactional relationships are being supplanted increasingly by strategic organizational mechanisms, such as sub-contracting and outsourcing, the evolution of managerial and social expectations may demand the fluidity that is represented in the diversity of FM operations—a function operating within a perpetual liminal state.

Adopting this stance allows FM practitioners to be viewed as active 'actors' as opposed to passive 'servants', that is, as service providers capable of casting their own 'spells' to co-construct the organizational realities they are charged with maintaining. The fluid nature of FM allows it to be present throughout different organizations, sectors, and hierarchical levels, with 'each actor of the FM business writing and proclaiming their own definition of facility management' (Nor et al. 2014). It is these 'hidden' (Becker 1990)—and oft-considered insignificant—roles that represent the rudiments of the narrative of FM as it seeks to metamorphose itself in order to compete with other secondary services in the corporate landscape. As a service provider, these identity mutations are placed within an increasingly captivated space, one where 'forms of enchantment will play an increasing role in shaping our understanding of what work is, and whose purpose it serves' (Endrissat et al. 2015: 1574).

The literature would therefore indicate that FM narratives are becoming refocused and remade within the storied spaces of the discipline. There is fluidity in the nature of the characters and roles that underscore the identity of the discipline (Section 2.1). A main association with the built environment is still evident, but the nature of the latter is also changing amidst the onset of 'third space' working (Jeffres et al. 2009, Okhuysen et al. 2013) and 'knomadic' knowledge work (Moravec 2013), both of which highlight the concept of liminality in the world of work generally (Section 2.2). Lastly, the strategic growth of outsourcing in defining FM processes and operations has highlighted the quest for occupational meaning in the world of secondary services—a world that increasingly positions FM as a metaphorical storyteller and enchanter (McCarroll 2017).

Therefore, with a view to the intangible nature of service delivery and surpassing the technical and predictable routines of maintaining the physical space of organization, we argue that FM represents a service positioned as benefiting from the skills and craft of storytelling. In evolutionary terms, we agree with Yang (2013: 135) that 'elaborate storytelling must have emerged as an adapted cognitive device for collecting and sharing important social and geographical information', with this practice being focally 'critical for our foraging ancestors in terms of the successful exploitation of scattered resources'. In this vein, the argument is made that storytelling represents an organizational survival mechanism, with the question of how this mechanism is represented in the evolution of FM being the exploratory focus of the next chapter.

3
A Tale of Tales

> When people told themselves their past with stories, explained their present with stories, foretold the future with stories, the best place by the fire was kept for . . . the Storyteller.
>
> **Jim Henson, 'The Storyteller' [television series, United States], Henson Associates, 28 April 1989**

Reflecting the essence of the Jim Henson quote (above), the research method detailed in this chapter focuses on the development of storytelling as a folkloric approach to organizational investigation. It is examined in line with research aim one—where the concept of the 'storied space' is examined; a concept which intertwines fiction and fact by utilizing the vehicle of the fairytale. In analysing the research potential of organizational storytelling, we develop a discussion that makes reference to wider issues of symbolism, representation, and allegory. This provides the reference point for a focus on the role of fairytales in discussing notions and claims of organizational enchantment and magic. Above all, the structure of fairytales will be examined in a way that allows story arcs to be analysed and character archetypes identified. This is achieved in line with research aim two, and thus discussion of key theoretical concepts such as liminality and metamorphosis.

Above, we noted Yang's (2013) argument that the craft of storytelling is a cognitive device attributed to helping humankind evolve—primarily a vehicle of knowledge transmission that reflects a normative coping mechanism. This point is embroidered in Fisher's (1984) well-quoted view that humans are essentially storytellers—*homo narrans*—with stories representing the communicative process whereby cognitive functions create meaning. This argument is strengthened when reflecting on the importance of storytelling for meaning-making in childhood years (Dietz and Silverman 2014). There are consequently many forms of storytelling and a range of typologies for analysing tales, legends, and sagas and their uses. However, storytelling and business management—the focus of this study—are often considered incongruous phenomena (Yang 2013), with the former conjuring images of fiction and fantasy while the latter of fact and control (Gabriel 2000).

Once Upon a Time in Facilities Management. Patricia McCarroll and John Hassard, Oxford University Press.
 DOI: 10.1093/9780191943492.003.0003

Moreover, the definition of what a story is has many variables (Hyde 2008) and can reflect various literary approaches and viewpoints (Jeyaraj 2019). Like FM, a universally accepted definition is elusive. Storytelling represents an ongoing arrangement of experiences—a sequence of events that is 'configured into a temporal unity by means of a plot' (Polkinghorne 1995: 5, as quoted in Oswick 2008: 141). It characterizes the evolution of experiences through the role of interpretation, rather than providing a simple documentary report on the past (Alvesson and Skoldberg 2000).

This effectively moves stories from being a relay of facts recounted to an enriched chronicle infused with meaning (Gabriel 1995, Hyde 2008)—one allowing for both real and imagined elements to be considered (Oswick 2008, Easterby et al. 2012). In this way, a narration—an ordering of events—is established that allows elements to appear that are otherwise hard to communicate; for stories can create a variety of realities, encouraging 'a landscape for dialogue and learning' (Pässilä et al. 2013: 160). At a broader level, it is sometimes argued that there is a practitioner-orientated approach associated with purposefully designed stories on organization (Denning 2001). The main function is to give real-world advice to managers/leaders, a practice often stimulated by a drive to facilitate transformational change (Parkin 2010), with stories such as these thereby providing 'sense-giving' tools (Collins 2013: 48).

An alternative approach branches from critical management studies (CMS), which carries with it a different focus—that of the complex, pluralistic, and poetic interpretation of stories. In this regard, narratives are viewed more as sense-making phenomena. Storytelling here highlights issues of power and transformation, and the importance of multi-focal and competing voices—factors and forces affiliated to this research and the aim of unlocking patterns of a disciplinal storied space (aim one).

Ultimately, this study sees stories as 'narratives—with plots and characters, generating emotion in narrator and audience, through a poetic elaboration of symbolic material. This material may be a product of fantasy or experience, including an experience of earlier narratives. Story plots entail conflicts, predicaments, trials, and crises which call for choices, decisions, actions and interactions' (Gabriel and Griffiths 2004: 115). The concept of pre-stories, or *antenarratives* (Boje 2011) is recognized, but is deferred in favour of a beginning, middle, and end (BME) format that allows the storied spaces to be overtly formed.

It is recognized that this would appear to be a structuralist approach, and as such to support the view that not all narratives are stories (Gabriel 2000). However, there is a firm recognition of the illogical events and fictional elements within this framework that allows for multiple interpretations and meanings, complex communications, and emotional performances. As such, the emphasis is not so much on recording simple facts, but rather in uncovering rich and enduring meanings, a point, which is emphasized in Section 3.1. The BME structure is then positioned as a means to contain stories, acting as a method to allow a conceptual reframing 'by selectively distilling disparate and often contradictory events and experiences into a coherent whole' (Rostron 2014: 97).

Organizational storytelling consequently is comprised of a myriad of different approaches and has been used in a variety of ways within organizational research. Gabriel and Connell (2010) emphasize this by positioning stories as an effective tool in the contemporary world of information overload. They help restore the balance of traditional knowledge and lore accumulated through experience with the facts of scientific knowledge (Gabriel 2000), with this association being linked to the concept of relationship management as discussed in Chapter 2.

With such a wide and evolving remit, many typologies have been developed to categorize the different types of stories and align them with their functions. The most long-standing of these is the work of Aristotle (365BC), who inspired Gabriel (2000), Booker (2004), and Hatch and Cunliffe (2006) in identifying basic classifications. These plot types represent the building blocks that formulate a story either as a dominant theme or a combination of premises. Appendix 1 illustrates that despite the myriad of stories created, they can be clustered into four main groups of *comic, tragic, romantic*, and *epic*. A *comic* tale is amusing but can also function to challenge organizational power and legitimacy (Gabriel 1991b). *Tragic* tales are rarer in comparison, and hold a cathartic function, which often absolves the protagonist from any responsibility in the face of oppression. *Romantic* tales reflect partnership and compassion. While the remaining category of *epic* tales are arguably the most visible within the organizational landscape (Hatch et al. 2005): 'If most tragic stories and a great many comic ones focus on *victimhood*, epic stories focus on *agency* and in particular on noble or heroic deeds' (Gabriel 1991b: 435, italics original). Taken simply, each story category has a main protagonist and a plot focus that aims to generate a specific emotion. Hatch and Cunliffe (2006) applied these in studying organizational functions (see Appendix 1), underlining why these specific types of stories may be told organically, and what the intended effect would be in terms of a deliberate message. In this way, the basic plots can act as a 'cognitive map assisting organizational sense-making' (Gabriel 2008: 282). This illustrates not only the role of storyteller but also the audience in the performance, with a successful story being 'the result of a triadic collusion between author, actor and audience' (Clark and Salaman 1996: 179).

This understanding of basic plot types and their overlapping categories supports what Martin et al. (1983, as cited in Boyce 1996) coined the 'uniqueness paradox', where each story is viewed as unique to the organization despite the potential contradictory nature of that reality. 'No matter how original our life stories may seem to us, we rarely are able to break away from a master narrative that sets the frame in which our personal stories are received, shaped and interpreted' (Zipes 2016: 277). Such tales can offer insight into what Gabriel terms the *unmanaged organization* (1991b), a space of spontaneity, freedom of expression, and creative experimentation (Lam et al. 2018)—one that has clear links to the notion of liminal space.

Narrative, story, and tale are used interchangeably throughout this research, although it is recognized that there is a distinction between them made in many studies (see Cunliffe et al. 2004). Based on Bruner (2002), and detailed in Søderberg (2003), a narrative has various attributes: first, narratives are spoken or written accounts of connected events occurring over time. This does not necessary indicate a chronological sequencing but is mainly reflected, in this study, by taking recourse to the generic BME format.

Second, narratives are retrospective interpretations of sequential events from a certain point of view. They essentially focus on a plot, where the storyteller prioritizes sequences from their own point of view, making the connections from their own understanding. Narratives contain characters, which carry the action and display qualities of human interaction. This allows the storyteller to reinforce or recreate their identity as well as of those they cast into a story.

The final attribute is that narratives are 'co-authored' by the audience (Hyde 2008). Walter Benjamin (1892–1940) highlighted this quality in his seminal work on storytelling (see Benjamin 1973). Working within the Frankfurt School tradition, he identified stories which are generated from 'afar', and those generated from 'home' (Hyde 2008). These represent two basic divisions—nestled within respective contexts, carrying distinctive characteristics, but offering considerable insight when weaved in combination by the oral powers of a storyteller. When narrative strains are separated, the storyteller is removed from the storytelling act (such as the solitary act of reading written text) and the stories themselves (stories no longer being formed through personal experience or by being passed from those who have had the experience). This results in information content surpassing the power of experience, the 'act' of the story fading in importance, and the role of the audience being detrimentally changed. The concept of experience is central to many of Benjamin's interests and has been the cause of considerable debate. Essentially, there is the *erfahrung*, a genuine affective experience, and *erlebnis*, 'a more immediate but disconnected experiential mode' (Latham 1999: 460). The lack of importance placed on experiences—or in '*having counsel*'—is connected to wisdom, and so the demise of storytelling has a wider implication for society, and ties into the disenchantment argument as presented in Section 2.3.

As seen in Chapter 2, FM represents a secondary service, one with an unclear identity and ambiguous functionality. It has a strong reliance on the parental organization (either internal or external) which affects any degree of permanence and arguably results in a state of liminality (illustrated in Figure 2.4). In this vein, FM aligns with management consultants who are storytellers, carrying the ability to create realities for their clients, capturing their imagination, and then, ultimately, their commitment (Clark and Salaman 1996) in contractual situations. These organizational stories carry with them functionality amid this chaotic uncertainty—as an attempt at restoring order. In the way that management consultants have been analysed to present the source of expert knowledge (Clark and Salaman 1996), FM can be positioned to 'shapeshift' into whatever is required (Shockley-Zalabak 2002), helped by its ambiguous nature. 'They tell tales about miraculous strategic virtuosity, of heroic organizational turn-rounds and of battles with organizational monsters (poor quality, poor service levels etc.). They tell tales about the necessary virtues for organizational success and how these virtues may be gained' (Clark and Salaman 1996: 177). As such, it is necessary at this juncture to explore the role of fiction within organizational storytelling, analysing the role of extended metaphor and symbolism, before focusing on the chosen genre of fairytales.

3.1 Mirror, Mirror, on the Wall

Organizational storytelling is therefore a well-recognized stream of meaning-making—one that is pervasive within the corporate landscape. The accepted definition of storytelling within this study has been declared as including the role of poetic elaboration and emotion (Gabriel and Griffiths 2004), which raises the question of the authenticity of a non-factual organizational tale. 'Acknowledging the existence and importance of narrative means that we need adequate devices for interpreting them, which leads us in turn to literary theory. Literary categories such as genres can serve as metaphors or even analogies, for other kinds of institutions' (Czarniawska 1997: 6). As the genre of fairytales has been declared as the main research approach within this study, the role of fiction in organizational research needs to now be explored.

Fiction loosely consists of three categories—non-fiction, narrative fiction, and semi-fiction (Yeung et al. 2016). Non-fiction offers highly factual accounts of actual perceived events. Accounts of narrative fiction represent works of the author's imagination and fall into two sub-groups, realistic and non-realistic. Realistic fiction may contain historical, accepted facts but is placed within an invented context, such as futurist sci-fi projections. Meanwhile, non-realistic fiction portrays elements that could not conceivably happen in real life; as such, they often contain reference to the supernatural (Yeung et al. 2016). Semi-fiction makes direct claims and reference to reality, although these may be embellished as it is positioned in a fictional setting. This requires a suspension of disbelief—effectively asking the reader to indulge in make-believe for dramatic communicative effect (Whiteman and Philips 2008). It is therefore apparent that there is much overlap between these categories, leading to no definitive boundary being identified between truth and fiction (Whiteman and Philips 2008), with hybrids of fact and fiction being common (see also Rippin 2011).

All of these forms have been utilized within social science, with the strength of fiction resting on their ability to be plausible to the wider audience; or contain an 'authentic' message, as associated with folklore in this study. This plausibility is emphasized in folklore using metaphor; with the definition of metaphor being 'two concepts of different things embodied in a single word or phrase . . . whose meaning results from their interactions' (Strati 1997: 309). Metaphoric power rests in that fact that these uses are sensed and felt (Drake and Lanahan 2007). This aids not only experience but also the ability to conceptualize (Kelemen 2000) across into the world of organizational life. 'The use of metaphor implies *a way of thinking* and *a way of seeing* that pervades how we understand our world' (Morgan 1998: 12, italics original).

Metaphor, however, can also be viewed as distancing and constraining (McCabe 2016), and so folklore's ability to also draw on symbolism (Trubshaw 2002) mitigates this recognized concern, supported by the assertion that organizations 'are by their very nature symbolic entities' (Pondy et al. 1983: 4). Symbolism denotes something larger than itself that requires an association, consciously or unconsciously, to create meaning and significance. Storytelling is therefore symbolic behaviour as 'the accounts are not the events themselves but representations of them created by the narrators and audience through a process of communication, interaction and feedback' (Jones 1996: 2), thus investing them with meaning: 'This is the essence of folklore: traditional, expressive behaviour generated by people as they interact with one another' (Jones 1996: 10).

The validity of using metaphor has been a cause of significant debate (see Tsoukas 1991), although the growing use of metaphor is illustrated for example through research by Harris et al. (2013); which creatively used the tale of *Cinderella* as an extended metaphor to explore gender advancement in academia. Metaphor used as a tool in organizational folklore is visible in Tangherlini's (2000) research into the working lives of paramedics, with Smith et al. (2014) also using folkloric principles to analyse issues of leadership and control in the police.

The use of extended metaphors as allegories is recognized in research (Höpfl 2000) but this is arguably rarer, methodologically, than metaphorical studies (Enninga and van der Lugt 2016). Allegory as a fictional mode of analysis is a means of writing 'other-discourse' (*allegoria*) (Cowan 1981: 113); a rhetorical device that operationalizes metaphors by creating a relationship between items within a story format (Tsoukas 1991). Practically it represents a two-fold process, where initially there is an understanding of the truth that is being presented before a subsequent recognition emerges of its absence. This is highlighted for example by Yeung et al. (2016), whose work within risk management allowed allegorical stories to be formed—ones that revealed 'the realistic "as is" mode and imaginative "as if" mode': a process 'lead[ing] participants from problem shifting to problem solving' (Pässilä et al. 2013: 160).

Consequently, this not only designates the presence of paradox in such literary devices, but also has repercussions for the concept of truth. In accord with work by Cowan (1981), truth becomes not something that can be possessed and presented, but rather part of an ongoing process that expedites its final depiction. It is this through representation and interpretation that using fiction adds value within organizational research (Hassard and Holliday 1998), in that the 'poetic truth' is more profound and subtle, whereas 'scientific truth' cannot provide this to the same effect (Arnoux and Soto-Andrade 2019). By creating stories that intertwine realities with fictional symbolism, the writer 'plans, thinks and calculates . . . the reader [then] does not respond or react to the words or sentences as physical objects, but understands their meanings' (Park 1982: 422), thereby symbolizing the creation of general set of values (Ashliman 2004). Benjamin (1968) also addressed allegories, not only in terms of the outward form of expression they portray, but also the inner experience (the intuition) they encompass—forces which transform the concept of allegory from being an 'illustrative technique' to a much more powerful 'form of expression' (Cowan 1981). A prominent example of this lies within the narrative of George Orwell's (1945) novel, *Animal Farm*, which has been positioned as a hypothesis of power and social structure in a manner similar to that developed in works of Max Weber (Phillips 1995).

As stated previously, storytelling conjures images of fiction and fantasy whereas management invokes metaphors of facts and control (Yang 2013), although the two are arguably interlinked in practice. The use of allegory is prominent within the growing sphere of 'reflective writing' (Bolton 2010), and Gabriel and Connell position case studies as representing a 'terrain somewhere between fiction and factual accuracy . . . trodden with permissible and acceptable variations' (2010: 510). Examples include Jermier's (1985) short-story explorations in radical organization theory and Fox's (1993; also 1992) ethnographic work in health care, where a fieldwork study of an orthopaedic surgical event was combined with Douglas's (1980) narrative fiction, *Bleeders Come First*, in order to evoke aspects of the experience that might have been missed 'if only the "true" event has been recounted' (see Bate 1997).

This fictional element of storytelling represents a 'narrative truth, where the validity of the story is made by the connection of the teller and the reader to the tale being told' (Hatch and Cunliffe 2006: 43), resulting in an engagement with the story's meaning as opposed to its 'hard facts' (Gabriel 2000). Therefore, fiction and its sub-genres are not fallacious to reality (McCabe 2013) but act rather as an extension of 'what has been extrapolated from the world we know', however much the illustrative narrative 'may soar beyond it' (Bruner 2002: 94). Allegorical stories, therefore, are symbolic reconstructions representing sense-making that provides further insight into the reality of situations and experiences in a manner driven by 'plausibility rather than accuracy' (Weick 1995: 55).

Consequently, it is apparent that the concept of *tellability* (van Hulst and Ybema 2020) is important, especially regarding organizational narrative where stories are often co-produced and can be 'so sensitive that they cannot be told truthfully or factually' (Smith et al. 2014: 222). The use of fictional storytelling can allow the teller to express things that may not otherwise be acceptable: 'poetic licence is the prerogative of storytelling ... we gain access to deeper organizational realities' (Gabriel and Griffiths 2004: 114). It is fitting to recognize the importance of authenticity and perception within organizational research as a recognition of fiction within organizational practice itself. For instance, Barry and Elmes (1993, as cited in (Küpers et al. 2013: 84)) go as far as to define 'the very phenomenon of organizational strategy as a form of fiction' as emphasizing that it is the amount of belief invested in strategic phenomena that represents the most significant aspect of their theory and practice. This is arguably supported by concepts such as lean management and agile management being positioned as 'fantasy constructions' (McCann et al. 2015: 1560). In terms of the discipline under study, Chapter 2 revealed this element in FM, focally in the role of FM professionals as service providers. Here, the role of client perceptions, rather than demonstrated 'hard facts' (such as key performance indicators) are emphasized and acknowledged, that is, acuities supporting areas such as relationship management and organizational dis/enchantment. In this respect, it is the client who is the narrator—someone who 'does not represent reality but creates a reality' (Abma 2003: 225), with expectations and perceptions being recognized within the industry as 'fuzzy' (British Standards Institution 2011).

This supports the analysis of Chapter 2, where this element of 'fiction' was visible externally in the role of service provision, but also internally, within FM, through the ambiguity associated with the practice. It is arguable that this presence of fiction can be extended across the organizational landscape: 'Notions such as identity and experience have lost their meaning; the only thing that applies is a façade, the front that is presented by organizations and toward them by their employees and managers. One has to look like a boss, present oneself like an experienced employee, make a good impression, and repeat it over and over again' (Kociatkiewicz and Kostera 2014: 13).

The storied spaces of FM represent a complex terrain, supporting the use of allegory to help explore this unique secondary service. FM illustrates role experience and perception dominating in a paradoxical balance for an industry dominated by measurements and outsourcing. The existence of different interpretations from one reality that can be paradoxical in nature is symbolic of the complex nature of social systems and therefore of storied spaces. This is a central quality of all complex systems (see Baskin 2008), made visible by the selection and sequencing of storytelling, and the use of metaphors that create an emergent meaning that was not readily available previously. In this manner, the use of metaphors is a vehicle of sense-making (Weick 1995), capturing 'how values and attitudes affect actions while symbolic acts influence beliefs, a fact not accounted for by a mechanical or machine model of organization' (Jones et al. 1988: 13). It is therefore projected that folklore, as allegory—and specifically the use of fairytales—represents an extended metaphor. It is this specific area that is now analysed in terms of defining folklore in relation to organizational studies.

3.2 Did You Ever Hear?

The definition of folklore, in line with storytelling and indeed the concept of FM, is also one of debate (Thompson 1946, Welsch 1968). Folklore is 'a hybridized discipline' (Bronner 2016: 7), incorporating anthropology, cultural studies, ethnic studies, gender studies, and history for example. This has been interpreted as a strength rather than a weakness: 'folkloric studies have become a battlefield on which scholars from various disciplines meet to dispute the theories and to contest interpretations. The resulting strife and dissension have proved surprisingly healthy and have fostered a spirit of vitality that otherwise might be absent' (Tatar 1987: 429).

A definition some decades ago from Ben-Amos (1971) places folklore as artistic communication within small groups. This definition does not, however, focus specifically on the oral tradition—synonymous customarily with the study of folklore—thus allowing for the inclusion, for example, of recent technologies of the digital age. This is illustrated from early works on Xerox/photocopy lore, as identified by Dundes and Pagter (1987), to the contemporary world of e-folklore (Krawczyk-Wasilewska 2006); which has created a new way of interpreting and communicating, representing a hybrid form of oral and literary creation (Hajduk-Nijakowska 2015). A more recent definition places folklore generically as 'traditional knowledge put into, and drawing from, practice' (Bronner 2016: 15).

Despite debates over definitions, folklore, at its core, is people-centric (Hallet and Karasek 1991), focusing on the *lore*—the learning—of ordinary people, and emerging from 'the folk', as opposed to being created and marketed by popular culture. It has close associations with tradition, and of the creation, transmission, and ultimately repetition of knowledge that is informal and unofficial, and yet remains consistent and distinctive to a group. It is the everyday, the unquestioned, and the taken for granted. It is 'those symbolic forms and process that are generated in people's . . . interactions that are repeated, emulated, or reproduced so as to become traditions or traditional' (Jones 1991: 193).

The *folk* of folklore represents the majority of people, in different societies, who define themselves through communal associations, whether they interact in groups of long or short duration (Zipes 2005). Traditionally folk are associated with 'oppressed classes' (Propp 1968), and with themes of protest and defiance against dominant classes or groups (Gabriel 2000)—themes which are transmitted orally by a largely illiterate people (Hallet and

Karasek 1991). The intended use of such themes was to explain, for example, natural phenomena (weather, seasons, etc.) or to celebrate recurrent rites (birth, marriage, etc.). As such, their function was to support communal harmony and stability, and produce tales of 'initiation, worship, warning and indoctrination' (Zipes 1994: 10).

Accordingly, these oral tales have no single identified author, but rather a 'series of narrators whose relationship to the tales is both intimate and detached; the folktale is the "extra-individual" . . . it exists both within and beyond each individual and personalized setting' (Benson 2003: 19). Literary fairytales, which are written down and traced to one author, appeared from the seventeenth century. There is no distinct line dividing oral and written traditions (Tatar 1987), but the emergence of literary tales changed and broadened the nature of the folk. Therefore, contemporary definitions see folk as a group that share at least one common factor (Dundes 1980). In this way, individuals can also be understood as 'part time folk' (Dundes 1980) to reflect how they move in and out of different folkloric discourses of social groupings, forming repeated expressions or practices (Bronner 2016) in each different storied spaces.

As such, folklore manifests in many forms. Dances, rituals, ceremonies, superstitions, traditions, and customs all fall within its scope. Organizationally this is vividly illustrated in Rippin's (2011b: 823) research on the symbolic use of ritualized head wear in office celebrations—characterized in her phrase: 'Pass me the tinsel, Mother: it's the office party tonight.' Additionally, in verbal form, folklore is represented in slang, jokes, jargon, proverbs, and stories (Gabriel and Griffiths 2004).

Folk narratives are divided into three main groups: folktales, myths, and legends. Folktales are 'an oral narrative form cultivated by non-literate and literate people to express the manner in which they perceived and perceive nature and their social order' (Zipes 1979: 7). As described above, fairytales fall within the category of folktales. The terms folklore and fairytales are used interchangeably within this study for ease of reference, although the fairytale is defined specifically by the inclusion of magic and a focus on ordinary working people, as opposed to gods and superheroes (Warner 2014a). Myth and legend arguably have more prominence within the organizational field, although there is much overlap. For clarification, a myth is 'a sacred narrative explaining how the world and humankind came to be in their present form' (Dundes 1997b: 5). McWhinney and Batista (1988) applied this to an organizational context in developing the notion of 'remythologizing'—identified as the process by which organizations both interpret and deploy their symbolism to enable renewal (see Boyce 1996). In contrast, a legend is defined as 'set in the modern-day world, frequently including real people and places and almost always describing a single event that often seems to transcend natural laws' (Trubshaw 2003: xvi).

Taken as a whole, organizational folklore as a concept first rose to prominence in the 1970s and 1980s, often being classified under labels of *Occupational Folklife* and *Industrial Lore* (Green 1978). Entwined in related disciplines such as anthropology, organizational studies, and sociology (Jones et al. 1988), examples can be found in journal publications of the time such as *Western Folklore* and *Journal of American Folklore*. '*Folklore* comes to be associated generally with the expressive dimensions of traditional culture; in contrast, *Folklife* commonly means the ways the group works together, and the devices deployed by the group in carrying out that work . . . their conjunction perhaps is more important than the distinction' (Abrahams 1978: 162, italics original).

Folklorists have long studied the customs and traditions—for instance, beliefs, songs, and tales—of occupations such as cowboys, miners, and roustabouts (oil-rig workers) (see Jones et al. 1988). During the 1970s, however, this scope was widened to encompass 'old folk in new locations and old relationships in new situations' (Green 1978: 221), such as the recognition of folklore within assembly lines and modern offices. Alan Dundes (1980) arguably pioneered this, with an 'American concept', which widened 'the European notions of peasant or class-based definitions [to include] repeated expressions or practices used folklorically within a family, locality, or occupation—or more temporary groups of friends, campmates, or music fans' (Bronner 2016: 9). This change in folklore coincided with a rise of the related field of organizational behaviour (OB), (Czarniawska-Joerges 1992), which has become established within the contemporary zeitgeist. Overlaps in the two approaches are clearly present: 'rarely does the word "folklore" crop up in the literature on work, labour relations, or management. However, with the rise in interest in organizational "culture" and "symbolism" over the past decade, more attention has been given to people's lore, particularly their stories' (Jones 1990: 279). This emphasizes the focus folklore places on the storyteller and the audience, 'permitting a view from the inside out rather than from the outside in' (Dundes 1980 xi).

When applied in an organizational context, folklore allows a gateway (Baruch 2009) into the storied space of FM, in line with OB principles: 'what recently has been called "symbolism" in organizations is "folklore", those expressive forms and communicative processes that have been documented and studied for centuries as "lore" and "traditions"' (Jones et al. 1988: 11). This is illustrated in a seminal paper by Rosen (1985)—'Breakfast at Spiro's'—which detailed the symbolism and social drama involved in a business breakfast; one which clearly revealed 'the relationship between cultural and social action in bureaucracy' (1985: 31). Gabriel (1998, 2000) also emphasizes this point, linking organizational narratives with earlier folkloric research approaches. Fairytales specifically, although relatively uncommon in organizational studies, are richly symbolic, containing familiar and habituated imagery with their use of plots and characters. In this manner, they offer a robust vehicle for a robust allegory—where both factors are easily recognized.

The view that organizational boundaries are increasingly in flux (Kociatkiewicz and Kostera 2014), coupled with the perspective of markets as being highly competitive, has led to organizations reflecting 'images of quest and trail' and hence in their actions 'struggling, like the hero, for survival against life-destroying forces' (Bowles 1989: 413). This is clear for example in the portrayal of CEOs in management bestsellers—such as *In Search of Excellence* (Peters and Waterman 1982) or *Good to Great* (Collins 2001)—which depict numerous heroic tales of executive success against the odds (see also McCabe 2009). Fairytales, with their specific inclusion of magic also allow a comprehensive storied space to emerge; albeit a space offering a freedom from trying to make sense of events in an orderly and controlled fashion, as often seen in modern organizational analysis.

The term *fairytale* itself is traced to France in 1697 when a collection was first published by Marie-Catherine d'Aulnoy. By 1750, the term was in common usage in England (Zipes 2011), but many stories associated with fairytales contain no actual fairies, as the name would suggest. Rather this is a link to the identifying characteristic of containing magic, leading to the fairytale also being referred to as a 'wonder tale': 'The character, setting and motifs are combined and varied according to specific functions to produce *wonder*. It is this sense of wonder that distinguished the wonder tales from other oral tales such as myth, the legend, the fable . . . wonder engenders astonishment' (Zipes 2007: 5, italics original). The verb to wonder also carries connotations of a receptive state—a suspension of disbelief found in fairytales, but also the active state of curiosity about the real (Warner 1994).

This is arguably why the fairytale is memorable and universal, with striking symbolism that endures in modern society. These tales represent an 'alternative' realm—to counteract the dictates of the dominating forces, historically manifesting as the Church or State, reflecting the common beliefs, desires, and fears of the people (Hallet and Karasek 2014). In this role, fairytales are enduring as they arbitrate between two worlds, creating a space for a dialogue to evolve that evades prevailing censorship. Always written in the past tense, fairytales utilize this space of ambiguity to be both retrospective and/or projective. The familiar, the ordinary, balances the element of magic, of wonder: 'the events that occur in fairytales are often unusual and most improbable, they are always presented as ordinary, something that could happen to you or me when out for a walk in the woods' (Gabriel 2000: 23). Fairytales represent both fortunate and unfortunate events (Zipes 2006) and can be used to stabilize or challenge (Zipes 1991)—positioned as a representation of liminality. As such, fairytales within the oral tradition were predominately aimed at an adult audience. This is illustrated through historical accounts of the philosopher Plato in the work of Socrates (470–699BC), who taught that fantastical tales should not be told to children, only to adults, as the former cannot judge what is allegorical and what is literal (Ashliman 2004).

The function, structure, symbolism, and archetypes of fairytales are the focus of this section of the study, in recognition of their 'evolutionary versatility' (Hallet and Karasek 2014: 23) as they move from the oral to literary traditions and into popular culture. Their format and simplistic characters allow 'windows to imaginative worlds inside that needed concrete expression outside in reality. They were to be shared and exchanged, used, and modified according to the needs of the tellers and listeners' (Zipes 2007: 2). Whereas the oral tradition represented communal values, the act of writing down fairytales not only changed the experience and voice of the tale to an individual one, but also made them the preserve of the social class that was literate.

The early literary tales in the seventeenth century were appropriated by the educated upper class, who used them not only for entertainment, but also as a sociological 'civilizing agent' (Zipes 1997: 3). An example of this is present in the works of French bureaucrat Charles Perrault (1628–1703), who in 1697 published eight stories in *The Tales and Stories of the Past with Morals* (*Histories, ou contes du temps passé*, subtitled *Tales of Mother Goose*). Intended for adults, *Sleeping Beauty*, *Little Red Riding Hood*, *Blue Beard*, *The Fairies*, and *Cinderella* are aimed at a female audience, with *Puss in Boots*, *Ricky of the Tuft*, and *Little Tom Thumb* aimed at a male audience. The gender dynamics of these tales portray the male role as necessary to temper the female, allowing her to behave appropriately and therefore achieve a 'happy ever after' state (Zipes 2006), applying this in the appropriate social context of the time, Perrault was arguably progressive in his views. 'Female characters are not mocked or parodied, Perrault wrote to illustrate what their "proper" place was in particular situations and how best to act in a civil way' (Carter 1977: xiv). For instance, Little Red Riding Hood, having no saviour and being unaware of how to behave in a dangerous situation, is consumed by the wolf. However, it is the submissive role of the female and the bravery of her male counterpart that set many of the standards associated with the modern fairytale genre.

These tales all had roots in oral traditions of the time. Perrault turned them into literary narratives from his perspective as a civil servant in the time of Louis XIV, who held a court of legendary extravagance at a time of significant hardship and famine for the French. Given the environment of pervasive economic instability (Carter 1977), Perrault turned the oral tales into parables, adding morals to the end of each tale to educate the reader; focally, to portray the way of the world and how to survive within it. The addition of morals, however, arguably undermines the effectiveness of the tales, making them appear frivolous or even cynical (Betts 2009: xxxv). Indeed, the tales often carried two morals, arguably to create contradictions and ambiguity wittingly. This ambiguity—the creation of space between worlds of fact and fiction—allowed Perrault to mock the superstitions of the folk of the oral tradition while also 'ridiculing the fashions and pretensions of the upper classes' (Carter 1977: xxii). The morals also provided distance from any elements of brutality contained within the tales, illustrating the sensitivities of the time. Perrault is also attributed with giving these tales their infamous symbolism, such as 'the red hood, the cat's boots, the glass slipper' (Betts 2009: xxv), with such imagery being reinterpreted widely in contemporary literature (see Carter 1977 on this issue).

The nineteenth century brought rationalization—largely affected by Christian influences—to the fairytale genre, as illustrated in the 1812 publication of *Children's and Household Tales* (*Kinder-und Hausmärchen*) in Germany by brothers Jacob and Wilhelm Grimm. This represented a much larger collection than the earlier works of Perrault (1697/ Betts 2009), including not only fairytales, but animal tales, legends, fables, and anecdotes (Table 3.1). The popularity of this collection is evident through the different issues released by the Grimm Brothers, who edited these tales across six other editions (Pullman 2012). Like the lasting contribution of Perrault, the Grimm Brothers' tales have been republished, re-imagined, and translated into 120 different languages (Zipes 2016) with the 1857 'final' edition being the 'second most popular and widely circulated book in Germany for over a century, second only to the Bible' (Zipes 1983: 54).

In contrast to the bourgeois audience of Perrault's court environment (1697), the Grimm Brothers' audience was arguably more working class. The brothers held professorships at the University of Göttingen specializing in linguistics and medieval studies, ultimately being dismissed for waging a formal protest against the King of Hanover over constitutional violations. As such, the tales they collected from 'servants, young women from middle-class families, housewives, watchmen and inhabitants from towns and small cities' (Zipes 1979: 6) held the intention to capture and preserve what the brothers saw as dying German traditions. However, as the editions progressed, the brothers introduced many changes to the tales, conceivably in response to criticism that they were too vulgar and unpolished for their class peers. The later editions, for example, are seen to downplay aspects of cruelty, with the addition of some Christian homilies (Zipes 2014a).

As displayed in the work of Perrault (1697) and the Grimm Brothers (1812–1857), therefore, the literary fairytale reflects not only the product of educated writers, but also that which tacitly mirrors personal beliefs and political ideologies of the time. The literary fairytale largely represents the 'bourgeoisification' of the fairytale genre (Zipes 1997), a condition where plural voices from the oral roots of the tales are replaced with rehabilitated narrative endings. It remains that the origins of fairytales 'presented the stark realities of power politics without disguising the violence and brutality of everyday life. Starvation of children, rape, corporeal punishment, ruthless exploitation—these are some of the conditions which are at the root of the folk tale, conditions which were so overwhelming that they demanded symbolic abstraction' (Zipes 1983: 8). Literary fairytales, clearly manifested across the forty-five-year span of the Grimm Brothers' publications, began to reflect therefore that era's requirement for a happier ending, ultimately forming the modern association with a 'fairytale' happy ending and in turn connection to a juvenile audience.

Indeed, further societal changes, such as the rise of the middle class from 1830 to 1900, concentrated images of fairytale endings being tailored to children (Zipes 2006). Across the globe, fairytales, with their familiar plots and enduring symbolism, come to form the initial literary and imaginative experience of children (Hallet and Karasek 1991). In a controversial study, however, psychologist Bruno Bettelheim (1976) analysed fairytales from a Freudian perspective, positioning them as an important symbolic vehicle for children to recognize and deal with traumatic events, extracting from Freudianism that the tales 'have a habit of translating phobias and fears into palpable physical presences rather than incarnating wishes and desires' (Tatar 1987: 71).

It remains, however, that modern fairytales are explicitly linked to childhood, with prevailing themes of innocence and utopia. This is highlighted through their connection with the works of Walt Disney (1901–1966). Although not explicitly a fairytale character, Mickey Mouse, the icon of Disney, has corresponding personal qualities—possessing no special talents but enjoying a personality that emphasizes sincerity and kindness in order to succeed (Ashliman 2004). The Disney Corporation represents the manifestation of contemporary fairytales to many—having created a global industry of 'fairytale' characters, films, and theme parks. However, this modern iteration of fairytales is not without controversy, in much the same way as the charges of vulgarity and violence levelled at Perrault (1697) and the Grimm Brothers (1812–1857) earlier. Given Disney is a multinational entertainment and media conglomerate—one underpinned by logics of mass consumption and shareholder value—this has led to arguments that Disney has translated fairytale narratives into media output not because the company 'believed their films had artistic merit and could contribute to children's cultural development' but more prosaically because it 'wanted to control children's aesthetic interests and consumer tastes' (Zipes 1997: 91).

The association with Disney films has also led to the related stereotyping of fairytale figures in modern society, where the term 'Disneyized' has been fashioned in order to define 'the saccharine, sexist and illusionary stereotypes of the Disney culture industry' (Zipes 1991: xxvii). This contrasts markedly with much of the socially utopian ethos of fairytales historically. To this extent, fairytales can become the 'popular trash . . . of imaginative literature', as debate rages between 'advocates [who] point to the way they ignite the imagination and spark hopes of liberty' and 'detractors [who] attack their suspect artistic quality and damaging . . . moral effects' (Warner 2014b).

The genre of fairytales—described in theoretical form in writing as early as Plato (Ashliman 2004, Teverson 2013)—has therefore evolved and changed across the ages and remains current in a variety of forms. 'Disseminated across a wide variety of media, ranging from opera and drama to cinema and advertising, fairy tales have become a vital part of our cultural capital' (Tatar 2002: xix). They are retold, re-imagined, and find new audiences through each decade. 'In a sense . . . folk/fairy tale has come full circle: originally intended for an adult audience, then censored and moralised for the benefit of children, now it is in the process of being returned to the adult reader/viewer' (Hallet and Karasek 2014: 20).

3.3 When Wishing Still Did Some Good

Fairytales therefore reflect a rich history not only in terms of chronological development but also in forms of socio-political message. They are 'reflections of social order in a given historical epoch' and as such 'symbolise the aspirations, needs, dreams and wishes of common people in a tribe, community, or society, either affirming the dominant social values and norms or revealing the necessity to change them' (Zipes 1979: 7). With magic being the defining element of fairytales, it is necessary therefore to analyse this in further depth, thus providing a link to the area of enchantment as discussed in the investigation of FM in Chapter 2.

In his pioneering work on mythology, James Fraser's *Golden Bough* (1890) placed magic as one of the human phases of belief, a stage superseded by religion and later science. This however remains controversial (Lang 1901), as significant areas of overlap and shared symbolism are apparent, as described in Weber's seminal 1917 lecture *Entzauberung der Welt*—the disenchantment or demagification of the world in the face of modernity (Davis 2012). 'In our modern society, our belief in and quest for functionality and rationality has come to mask the fact that our institutions are founded on a socially constructed, and therefore imagery, version of "world order"' (St-James et al. 2011: 634).

The oral tradition of fairytales used magic to help explain the world, provide emotional empowerment, and inspire solutions (Davies 2012). In the pre-industrial age 'magic was omnipresent, but not omnipotent' in that its effects could be 'harmful or beneficial' (Ashliman 2004: 15). It manifested itself in fairytales through the presupposed actions of both supernatural supporters and opponents. In the tales, magic was rooted in the world as an everyday occurrence, for example, in beliefs that it was not unusual for animals to talk.

However, magic also comes with rules that must be followed, such as in the tale of *Cinderella* (ATU 510), who must return by midnight, when the spell breaks. The role of magic remains essential and enduring, contributing to the fascination that is still current within the genre of fairytales. 'Although the knowledge involved may seem fantastic, it has emanated from some human experience, and whatever the experience may have been before it was imaginatively appropriated by tellers, who have given it structure and sense, we may be sure that the experience, as human, is significant' (Zipes 2016: 277).

Additionally, the concept of magic has multiple meanings, on the one hand carrying connotations of superstition and illusion, whilst on the other representing a rhetorical tool to understand relationships, with elements of this duality being found in the concept of enchantment (Davies 2012). Rather than representing a simplistic and one-dimensional belief in mysterious forces, St-James et al. (2011) argue that magical thinking allows for different positions toward the possible to be created. This 'maintains ambiguity around what is possible in order to provide meaning and sustain hope in the context of stressful situations' (St-James et al. 2011: 647).

Magic is also associated with the more physical attributes of modern technology. As Arthur C. Clarke (1968) once noted 'any smoothly functioning technology gives the appearance of magic' (see Shull 2005: 69). The magic of technology, presented in forms such as the worldwide web—which connects the globe in a seemingly invisible manner—emphasizes qualities of wish fulfilment and emancipation through its means of communication (Ashliman 2004), whilst also suggesting untapped potential and undiscovered capacities in this fast-paced industry. 'Along with the development of the internet, the computer has become a liminal, mystic intermediary that allows crossing the limits of personal perception and entry to the extra-natural virtual universe where everything seems to be viable' (Meder 2001 as quoted in Hajduk-Nijakowska 2015: 162).

In this manner, the idea of magic is linked to the ability to suspend belief and challenge norms, using imagination and creativity, qualities often associated with organizational demands for innovation. Concepts of magic in fairytales allow plotlines to evolve which are visible in the organizational 'epic' (see Appendix 1); stories that, for example, create a space to 'break old patterns in order to test the borders of the possible, explore new grounds and perhaps establish new institutions' (Kostera 2006: 13).

The presence of magic within fairytales also creates a space for the *other* or *otherness*: 'they open a door on Other Time, and if we pass though . . . we stand outside our own time, outside Time itself, maybe' (Tolkien 1947: 11). Although J. R. R. Tolkien's essay *On Fairy-Stories* is much debated (see Zipes 1979), he identifies a range of related factors that complement the concept of liminality—notably fantasy, escape, and recovery. Here he sees fantasy as a rational and normal human activity, something which represents a liberating force for psychological escape (Bærenholdt 2016). Recovery meanwhile is the ability to move into a different world to 'enable readers to regain a clear view of their situations' (Zipes 1979: 162) and recognize connections between past and present that allow a sense of renewal.

Magic, therefore, permits capacity for the *other*. It creates 'sacred thresholds in space and time which symbolize spiritual as well as social boundaries, and the transition across them' (Davies 2012: 19). According to Zipes (2006), it reinforces an idealism of change that holds universal audience appeal. With reference to the BME format of the fairytale, this allows for the fundamental fairytale concept of magic to be mapped on to our earlier analysis of FM liminality (as illustrated in Figure 3.1).

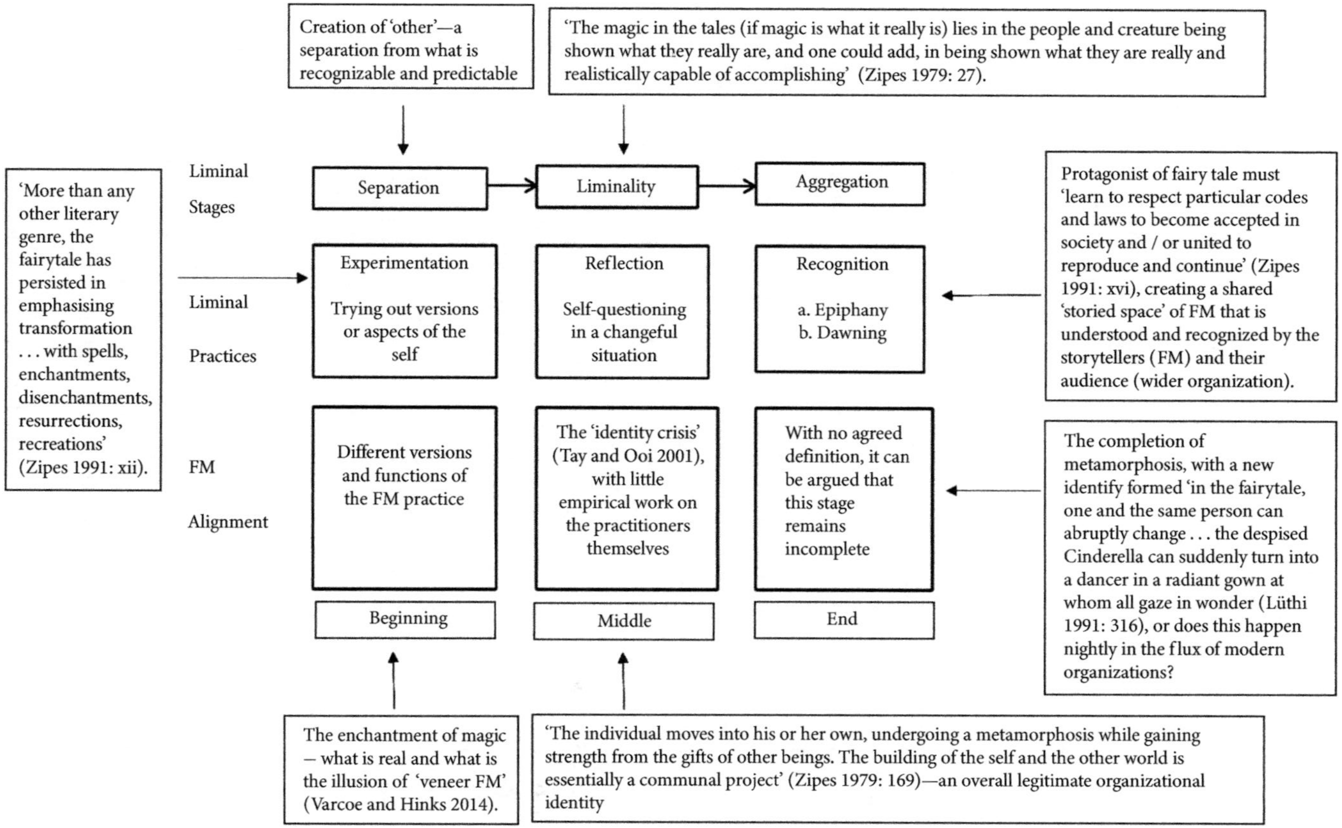

Figure 3.1 Magic and liminality
Source: adapted from Beech (2011: 290).

Magic allows for a separation from normality and a simultaneous questioning of current identity, with the removal of previous boundaries sanctioning different possibilities—both positive and negative—to be considered. This element of enchantment underpins the metamorphic power of the fairytale genre, where magic gives expression to thought experiments; for example, the wicked fairy turning out to be capable of love (Warner 2014b). With the pace of organizational life 'quickening to dizzying speeds heretofore only the stuff of dark fantasy' (Tyler 2011: 143), there is a strong argument that the separation stage is apparent not only at a secondary service level—as reflected in the function and operations of FM—but also in terms of broader organizational analysis.

The contractual nature of FM, as illustrated in Chapter 2, makes this process more salient: different pressures within the liminal space pressuring the new acceptable identity that provides the most symmetry with the parental organization, allowing aggregation. This could take the form of a *supernatural opponent*, who promises everything with magical illusion, as seen in the identification of 'veneer FM' (Varcoe and Hinks 2014a), or as a *supernatural helper*, which aligns to the overall mission statement of FM (IFMA 2022). This is the state of instability that the literature analysis of FM reflects, presenting a potential storied space that connects with the story arcs of fairytales—phenomena which have a 'penchant for moving from one extreme to another'; a situation 'vital to understanding . . . characters, plots and thematic orientation', for 'fairytale figures have very few fixed traits; they are reformed once they reach the goals of their journeys, when they become endowed with the very qualities in which they were once found wanting' (Tatar 1987: 102).

It is within this space that the element of enchantment, of magic, and therefore disenchantment is more perceptible, that is, whether FM can create an identity synonymous with magical problem-solving capabilities, such as reflected in the opening quote in this book, or succumbs to the spectacle of illusion, of false promises. 'You'll keep up the smoke and mirrors act because they are the only thing keeping the Emperor in new clothes' (Heath 2016, online). With the liminal process creating this space for reflection, the outcome is therefore in the storied spaces of this process—capturing 'the lessons of experience (either historical or mythic) and through their repeated retellings, reinforce[ing] conventions of appropriate behaviour and special norms' (Brewer and Dourish 2008: 967).

It is this search for accepted norms and behaviours to legitimize FM that complements another key trait of fairytales, that is, the concept of 'lack'. As Zipes (2007: vii) notes, 'though the fairytale is frequently emptied of its substantial utopian longings through commodification, it still denotes a loss or a lack'. This is a prime mover of fairytales, where lack can be seen in the absence of good, which creates villainy and/or a desire to have something (Tatar 1987). This ultimately leads to the notion of search; the protagonist's journey that reflects the fairytale motifs of combat with an adversary, a pursuit, a return. In support of the FM analysis of Chapter 1, the liminal space is evident in its search for an identity, reflecting both empowerment and alienation (Driver 2014).

3.4 There Once Lived

With core principles of magic and lack being identified with the genre of fairytales, there are also other characteristics associated with the field that have arguably sustained the evolution and plasticity of fairytales. 'Often disguised, re-arranged, changed, the Wonder Tale appeared to have a definite shape, and tended to be quite widespread, to be at home, as it were, in a large number of communities all over the planet' (Gillet 1998: 131). In the main, for example, there are few references to machines, technology, or other signs of industrialization, reflecting the predominantly oral roots of many of the tales. Additionally, the element of magic allows the concept of time to be fluid and unidentified, while characters rarely are named but instead identified by their occupation, status within a family, social position, or items they wear (Carter 1977, Pullman 2012). By omitting extraneous details, fairytales are also notoriously short in form, with the adventure itself being the message (Ashliman 2004). 'The notorious flatness of the fairy-tale character may discourage critics from trying to read their minds and analyse their motivations, but the boldness of their deeds invites careful scrutiny. Those deeds may be routed in realistic situations, but they are often so far removed from reality that they demand symbolic readings' (Tatar 1987: 56).

The simplicity of fairytales is therefore at a surface level, presenting clear and linear story lines, but with striking imagery and ambiguity that holds many meanings on different levels. This has enabled the genre to remain popular to a multitude of audiences throughout time—audiences interpreting a range of events which appeal directly to them. Narrative familiarity is found within every element—the plot, the characters, the images, and the motifs. It is this generic understanding that facilitates the flexibility and re-imagining of the fairytale and thus has ensured its enduring applicability. 'Fairytales thrive on simplification, focusing on polar opposites rather than on the complex continuum that connects them' (Ashliman 2004: 7). This moves the genre of fairytales from what Geertz (1973) coined 'thin' description, where events are recounted without much further investigation, to 'thick' descriptions, which seek to uncover meanings assigned by the narrator, the subject involved in them (see Drake and Lanahan 2007).

In this way, fairytales represent the 'concealed revealed' (Houlbrook 2018)—an ending that is not really an ending but rather another beginning, offering a sense of circularity, uncertainty, and anticipation, as opposed to the stereotypical happy-ever-after motif influenced by modern works and as associated, for example, with Disney productions, as analysed previously. Rather the happy ending is not the finale of the tale, but a prophecy of the future—one signalling the beginnings of a larger story (Warner 1994). Several tales have no conclusive ending, probably the best-known of these being the Grimm Brothers' fairytale *The Golden Key*. This tale details the discovery of a key and an iron casket by a poor boy, ending with 'so he turned the key around once, and now we must wait until he unlocks the casket completely. That's when we'll see what's lying inside' (Zipes 2014a). This was the last tale in all the versions of the Grimm Brothers' collections, starting from the second 1815 edition, 'suggesting perhaps that there are more marvellous tales yet to be discovered' (Pullman 2012: 420).

It is evident, therefore, that fairytales contain key motifs—ones sufficiently striking to be repeated, and consequently form part of a tradition (Georges 1997). The names and attributes of the characters change, but their actions and functions remain decidedly consistent allowing the tales to be recycled, re-imagined, and updated (Ashliman 2004). These traits reflect the central elements of narrative, which the storytellers feel obliged to repeat recurrently for listeners. In this sense, however, an act of narrative production cannot always ensure an equivalent act of narrative consumption, especially in terms of created meaning (Hassard and Kelemen 2002). As Lindahl (1997: 66) argues, 'a folklore plot may keep its general shape through hundreds of tellings yet mean many different things each time it is told; the stability of a pattern offers no guarantee of stable meaning.' Nonetheless, the plot of the tale typically addresses conflict, crisis, and resolution, and forces basic characters to be identified. This allows the protagonist to be established and roles for the cast to be allocated; for example, who is speaking and who is watching the scene, who is the change agent and who/what is to be changed etc. (Rotmann et al. 2015). The generic BME format also propels the narrative towards some form of culmination, be it recognition of a point for continuation or a final moral lesson to be reflected upon.

Fairytales in this light represent 'a vector of social memory' (Uther 1997: 260). However, as principal motifs demonstrate, they also guide actions present and future, selecting what the members of a community might find relevant (Zipes 2006). Placed into an organizational context, it is through work-based stories that participants predominantly communicate their experiences. This shapes our understanding at both individual and collective levels and thus helps form attitudes and identity (Gabriel 1995). With an employee's identity affecting how they undertake their role, the storied space allows a window into these 'strong plots' (Butler et al. 2014), which 'have been institutionalized, repeated though the centuries and are well rehearsed with different audiences' (Czarniawska 2004b: viii).

3.5 When We Get to the End of the Story

In response to the great number of tales collected in the nineteenth and early twentieth century, the Folklore Society was formed in the United Kingdom in 1878 (Simpson and Roud 2003). This followed the approach favoured at the time to record accurately as many folktales as possible, noting the date of collection, storyteller details, and so on in what was known variously as the 'comparative', 'geographical-historic', or 'Finnish School' method (after the nationality of some of its most prestigious practitioners). A by-product of this movement was the formulation of a generally accepted system of classification, *The Types of the Folktale*, first published in 1910 by Finnish folklorist Antti Aarne. This was employed as a base document, which was revised and enlarged, and provided a framework for Stith Thompson's six-volume *Motif-Index of Folk-Literature* (Thompson 1946). This index sought to dismantle individual European-based tales into constituent parts or motifs, thus detaching the tales from their original context (see Benson 2003). In 2004 this was updated, and once again enlarged, by Hans Jörg Uther to form what is now known as the Aarne-Thompson-Uther (ATU) index, cataloguing over two thousand tales, in numerical order, and subdivided into seven categories (plus 'unclassified tales'), as illustrated in Table 3.1 (see Ashliman 2004, Carrassi 2016). Fairytales, as the largest category, were divided into a further seven sub-categories (plus 'other stories').

The nature of folktales, however, is extremely varied and arguably does not easily lend itself to such clear and precise classification (as indicated above), a situation which has led to much criticism by writers such as Holbek (1987) and Zipes (2006). Nevertheless, the ATU classification system is extensive, detailing basic plots indigenous to Europe and the Near East. It thus remains a major reference point 'representing the keystones for the comparative method in folkloristics, a method which despite postmodern naysayers

Table 3.1 Aarne-Thompson-Uther (ATU) classification system

Group	Type	
Animal Tales/fables	1–299	
Fairytales / magic tales	300–749	300–399: supernatural opponents 400–459: supernatural relatives (400–424 wife, 425–449 husband, 450–459 sibling) 460–499: supernatural tasks 500–599: supernatural helpers 560–649: magic items 650–699 supernatural power or knowledge 700–749: other stories of the supernatural
Religious tales	750–849	
Realistic tales	850–999	
Tales of the stupid ogre	1000–1199	
Anecdotes and jests	1200–1999	
Formula tales	2000–2399	
Unclassified tales	2400–2499	

Source: Ashliman 2004

and other prophets of gloom continues to be the hallmark of international folkloristics' (Dundes 1997a: 2000).

Where the ATU catalogued story types, another seminal classificatory study was provided by Soviet folklorist and scholar Vladimir Propp (1968). Analysing more than a hundred Russian folktales in his publication *Morphology of the Folktale*, he identified 'an alternative, strictly defined set of invariable functions, which occur, to varying degrees, in all the tales and which point to a structural rather than a thematic unity underlying the surface multiformity' (Benson 2003: 26). This scheme to identify the simplest, irreducible, structural units of folktales results in thirty-one functions (absentation, interdiction, trickery, complicity, etc.) that represent key characters in terms of their position and role in the plot. The functions relate to specific roles characters may play in the tales, although not every function appears in every story, and in some tales a group of characters, acting together, may complete some roles (Fisseni et al. 2014). In the main, however, the functions do follow a sequential order, with exceptions—for example, literary signatures such as the 'trebling' (repetition) of 'nested structures' (see Fell 1977)—being rare, with no tale containing additional functions. Some of the labels used to distinguish functions however have been criticized, such as a wedding implying a ceremony when it can also imply a primary function of monetary reward (see Dundes 1997a). Propp's functional approach has been criticized for its excessive formalism—a criticism often aimed at Soviet narrative analysis—with one of his most ardent critics in this respect being the renowned French structuralist anthropologist and ethnologist Claude Lévi-Strauss (see Shishkoff 1976). In fairness, though, Propp's system is complex because it reflects the wide-ranging literary genre it is trying to encapsulate.

Extrapolating from the basic functions, Propp's analysis then identifies seven major character functions, or 'spheres of action': the *hero*, the *villain*, the *princess* (or *prize*), the *dispatcher*, the *donor*, the *helper*, and *the false hero*. As a tale unfolds, one character may have multiple roles, and several characters can be attributed to a single sphere of action (see Tatar 1987). In this way, Propp's work offers a formula-driven structure for making sense of the often-capricious realm of fairytales. This morphology has been used in many forms since its inception, from Alan Dundes (1980) folkloric work with North American Indians to more recently cyber gamemakers analysing Japanese folklore (Wama and Nakatsu 2008), the development of artificial life and robotics (Imabuchi and Ogata 2014), and literary and linguistic methods in computing (Fisseni et al. 2014). There is also a precedent for linking Propp's work to motion pictures in order to analyse 'westerns, detective stories, horror, science fiction and simple adventure tales' (Fell 1977: 20).

Around the mid-1950s, when Propp's work was first published in English (being originally published in Russian in 1928 but not translated until 1958), Claude Lévi-Strauss (1955) also sought to investigate a proposed structure of narrative form, specifically using myths and mythologies, and famously tracing the cultural evolution of a single myth from the tip of South America—and all of its variations from group to group—north through Central America and eventually into the Arctic Circle (see Benson 2003). He accomplished this by examining the underlying structure of relationships among the elements of the story rather than by focusing on the content of the story itself. Unlike Propp, Lévi-Strauss's method was algebraic, involving taking stories out of linear sequences and regrouping them to highlight structural 'paradigms', principally concentrating on conceptual opposites to discover underlying metaphorical meaning.

The structural approaches of Propp and Lévi-Strauss, however, are both susceptible to criticism as they remove context from the tales and myths analysed, basically reducing them to formulas and thereby creating 'folklore without the folk' (Dundes 1980: 33). A bridge between them, is arguably found in the work of Algirdas Greimas (1966) who refers to the folktale as a 'micro-universe', and one which 'exhibits the same structural characteristic that are common to the world as it is perceived by the human subject' (see Benson 2003: 32). Developing Propp's seven spheres of action in concert with the binary theory of Lévi-Strauss, Greimas identified the 'actantial model' (see Figure 3.2). This balanced dualities across six 'actants', defined as a structural unit of a function (Søderberg 2003), and which theoretically, in nature or form, can be human, animal, object, or emotion (Robichaud 2003). Placed as three pairs of binary opposites, the model explains fundamental narrative patterns along three binary dimensions, with *subject/object*

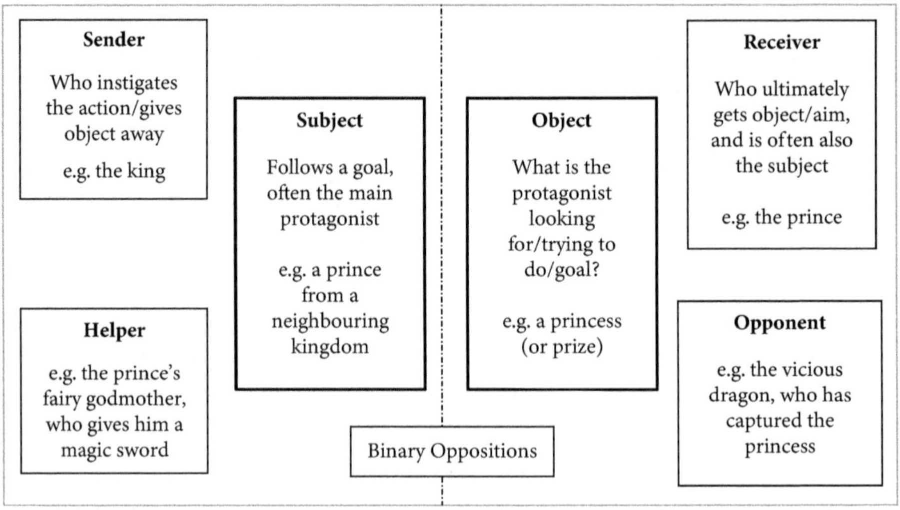

Figure 3.2 The actantial model
Source: adapted from Søderberg (2003) and Robichaud (2003).

representing the 'axis of desire': *sender/receiver* the 'axis of transmission', and *helper/opponent* the 'axis of power'.

Placing the above example into symbolic context, for the first binary opposition, the *subject* actant (a prince) is following a goal, charged with a mission, and associated to the main protagonist role, while the *object* actant, the goal or prize, is represented by a princess who is being pursued by the prince. For the second binary opposition, the *sender* actant (also referred to as the *donor*) instigates the action (which often mirrors an agenda of the sender—such as the king directing the prince to rescue the princess), while the *receiver* actant (often one and the same with the *subject*) represents whoever benefits from the aim (in this case, the prince who 'wins' the princess). The final binary pairing, of *helper* and *opponent*, sees these actants either assist or hinder in achieving the aim (such as a fairy godmother who aids the prince in defeating a dragon which is imprisoning the princess). The aim of actantial model therefore is both to capture forms of action/interaction and illustrate (actant) relationships in a symbolic network (see Robichaud 2003).

However, although the actantial model captures both the nature of characters and states of being—thus answering criticism of previous models—limitations have also been identified. For instance, the sender–receiver binary has been deemed problematic, for as Tatar (1987) argues, contextual detail can be eliminated from a narrative when it overly concentrates on the driver of a plot in the form of a goal. Nevertheless, the actantial model remains a well-used tool within narrative structures, allowing a surface-level interpretation to be made visible, especially in complex narrative structures (Søderberg 2003).

There have of course been many attempts over the decades to classify narrative forms and identify core plot structures. Discussing the power of binary oppositions, and in relation to the Propp versus Lévi-Strauss debate, Dundes (1997: 7) suggests they are 'just as strong in [analysing] folktales as they are in myths' (1997: 7), with this being illustrated by Lévi-Strauss seemingly evaluating myth and folktales wholesale; even though sometimes claiming to be studying myth exclusively. Dundes explains this argument by asserting that Oedipus is really a folktale, rather than a myth. Similarly, Lacey (2000: 65) argues that binary opposites are germane to such narrative analysis because folktales are structured by 'the attempt to resolve conflict, characterized by the opposition between the hero and villain'. However, for Lacey, one side of the opposition always holds an advantaged position over the other, and as such the hero, who signifies what humanity holds as right, good, or moral, 'invariably wins'.

Elsewhere, in studying the 'fantastic, the fantastic uncanny, and the fantastic marvellous', Tzvetan Todorov proposed a model, which, in terms of a more all-encompassing applicability, aims to expand on Propp and Lévi-Strauss and move closer to a generic prototype for narrative analysis. Todorov's theory comprises five stages: initial situation, disruption, recognition of disruption, attempt at resolution of disruption, and resolution, also understandable as a transformation from equilibrium through disequilibrium and back to (a modified) equilibrium. In developing this analysis of narrative stages Todorov posits eight basic plot lines, these being: *the quest, redemption, journey to another world, the beast transformed by love, the solving of riddles, the 'biter-bit', the stranger saviour*, and *rise and fall* (see Lacey 2000). In many ways, this model can be mapped onto Propp's functions, although due to Todorov's classifications being more flexible they can be applied to a far wider array of narratives.

More influential, and certainly with far greater popular profile as a theory of narrative and plot structures, has been the method developed by Christopher Booker (2004), who identified seven plots common to all stories (see Appendix 1). Influenced in particular by the works of psychoanalyst Carl Jung (1875–1961), and in a book that Booker worked on for over thirty years, he identifies plots which have extensive areas of overlap, namely, *overcoming the monster, rags to riches, the quest, voyage and return, comedy, tragedy*, and *rebirth*. Given the profile and influence of Booker's work we will briefly give space to outlining the basics of these seven plots and principally in relation to the ATU classification.

Overcoming the monster corresponds to the ATU *supernatural opponents* (300–399) category. In these tales, the focus is on the monster and how the protagonist confronts it. 'The monster is heartless: totally unable to feel for others. Although this may sometimes be disguised beneath a deceptively charming, kindly or solicitous exterior: its only real concern is to look after its own interests, at the expense of everyone else in the world' (Booker 2004: 33). It is the outwardly caring but evil stepmother, the witch in *Hansel and Gretel* (ATU 327). The monster's view of the world is warped by both its own egocentrism and limited understanding of the world in which it exists, ultimately culminating in a state of vulnerability that is the key to the monster's downfall. This is generally at the hands of the protagonist, who must defeat the monster to save the community/kingdom/world/prize and therefore generally be rewarded. The focus of the tale is identifying with the protagonist as they battle this seemingly unwinnable evil, experiencing their tension and fear, but also freedom and empowerment alternatively.

Booker's 2004 *rags to riches* classification does not map as precisely into the ATU categories as *overcoming the monster*, although it does have strong parallels with the *supernatural helper* category. This is arguably the most dominant plot associated with modern fairytales such as *Cinderella* (ATU 510), albeit with a focus on sentimentality and romance. The primary motif of these tales is that 'someone who has seemed to the world quite commonplace is dramatically shown to have been hiding the potential for a second, much more exceptional self within' (Booker 2004: 52). It is this moment of transformation, of reveal, that holds a profound appeal. These are tales of survival against those people who are blind to their abilities and who often seek to repress the protagonist, but ultimately, they represent the protagonist's journey to maturity. As such, the overlap with the *overcoming the monster* category is apparent, but here the focus is on the main characters 'being recognized for what they truly are. They must prove themselves worthy to move from a state of misery and poverty 'to a state of splendour and happiness' (Booker 2004: 65). For instance, it is not the fine clothes that Cinderella wears to the ball that form the lasting impression, but her innocence, inner beauty, and unpretentiousness (Zipes 2000), balanced with her ability and humility to follow the rules of magic. If the protagonist fails in this regard the narrative falls into the *tragedy* category.

Quest tales start with a call that causes a journey, with the important characteristic that the hero is rarely alone but has companions, helpers, who assist in sustaining the hero throughout the trials they face. The latter manifest, for example, in monsters, entrapment, and temptation: ordeals the hero must endure to succeed. The motif of *voyage and return* is the counterpart to *the quest*, with the protagonist, for example, overcoming threats in a strange land before coming home with riches (sometimes in the form of experience, such as in *Goldilocks*, ATU 171). In these tales, the emphasis is on a psychological journey, of a coming of age, of being in unfamiliar or alien circumstances, cut off from all things normal, which is at first exhilarating as well as confusing. As the plot develops, the themes of being trapped or threatened grows, and the protagonist must fight their way to return to normality and safety. This plot line arguably has the clearest synergy to liminality, with its focus on the psychological development of the main character. 'They have escaped from their original state of limited consciousness and learned to "see whole". They have discovered who they are. They have grown up' (Booker 2004: 223).

Booker (2004) maps *comedy* with *romantic* storylines, and in so doing often with motifs of growing, absurd conflict, and confusion, factors which often culminate in a single event. This category, along with *tragedy*, where the protagonist's own actions often result in their ruin, are arguably rare as central characters, but do appear in less-known tales such as 'How Some Children Played with Slaughtering' (Grimm Brothers 1812, as translated by Zipes 2014a). The last main plot according to Booker (2004) is that of *rebirth*, where the protagonist encounters an event or dark power that makes them change perspective, often becoming a better person because of it, such as in Beauty and the Beast (ATU 425). As with other characterization and classification schemes discussed, the seven plots can be used to reveal the hidden narrative structures of fairytales. However, in Booker's work the starting theme for all seven plots is universal, that is, the protagonist being in some way incomplete, which he identifies as an archetypal pattern.

With the ATU and Propp's analysis showing similar tales existing across the globe, the question of how this came about led to the theory of diffusion, where tales were narratively conveyed in the migrations of the storytellers, picking up cultural variations along the way. In contrast, the theory of fairytales displaying an archetypal pattern lends itself to the theory of polygenesis, where the 'sameness of the human psyche led to independent invention of stories with similar plots in two or more regions' (Ashliman 2004: 16). The travelling storytellers of the oral traditions no doubt helped spread the tales, but polygenesis indicates the presence of a deeper level of unconsciousness that exists in genetic inheritance. This is central to the psychology of Carl Jung, and in particular his 1945 thesis '*Zur Psychologie des Geistes*' (On the Psychology of the Spirit), which frequently cites fairytales in understanding the stable state and working of the human consciousness. 'As Jung observed, the content of fairytales is not subject to the whims and caprices of individual logic but is determined by eternally valid processes of thought' (Tatar 1987: 81). In Jungian theory, humans have evolved a 'collective unconscious'; one revealed though the presence of archetypes, which represent the unconscious context that is altered by transfer into individual awareness (Jung 1969b). Holding dual properties of negative and positive (Drake 1969), archetypes are expressible only through culture and its meaning making (Hatch et al. 2005), and as such possess the ability to unite individuals into collectives (Kostera and Obłój 2010). Archetypes are therefore 'the riverbeds along which the currents of psychic life have always flowed' (Jung 1964: 227), thus representing the hidden knowledge and values which provide meaning, through pattern formation, with which to help us navigate the world.

Fairytales are therefore archetypal; seen in oral narratives the world over and representing universal symbols of good and evil, light and dark, strength and weakness (Schneider and Shrivastaval 1998). Although open to much interpretation and reinterpretation globally, such variation does not affect fairytales maintaining their basic structural form and symbolic character (Hatch et al. 2005). 'The king, the princess, a witch . . . are . . . for the human imagination, age-old symbols for what is high, noble, and pure or dangerous . . . what is genuine or true, or what is sordid or false' (Lüthi 1991: 317). As reflected in fairytales, archetypes act as themes that bridge the conscious and unconscious mind (Lewis 2011), by inspiring or disturbing what was previously unconsidered (Kostera and Obłój 2010). Archetypes represent powerful drivers in determining how ideas are formed and reformed, forces that guide human action. Where stereotypes are limiting, preventing emotion and reinterpretation, archetypes are both functional and moral, linking the collective knowledge and individual experience required for appreciating shared understanding in networks of meaning, such as those that organizations represent (Smircich 1983).

As we have argued, Chapter 2 positioned FM, a secondary service, as organizationally one of the conduits of this meaning, with this research aiming to target the archetypes and storied spaces at the heart of the performing of FM's functional role. Applying the concept of archetypes to an organizational setting enables hidden facets of corporate life to be revealed in a personalized form, thus positioning the corporate unseen/ignored with everyday experience in a research inquiry. The metaphoric element present in fairytales can be deployed to explore the imaginative aspect of a discipline (Kociatkiewicz and Kostera 2012). For example, as an approach to understanding the 'deep roles' of work organization, Moxnes (1999a, 1999b; also Moxnes and Moxnes 2016) expanded Jungian theory, mapping it with the familiar characters of fairytales as the basis for organizational analysis. Limited in number and emotionally charged, roles are filled with dualistic (good and bad) characteristics, reflecting Greimas's actantial model (Figure 3.2), with each actant potentially comprising several actors. For instance, a 'bad' *father* role can take many forms (*husband*, *devil*, *boss*, etc.) but consistently represents something to be feared (Moxnes 1999b). In this dualistic analysis, the primary roles are all family based (father, mother, son, daughter), with secondary roles being in the form of *helpers* (spiritual helper and material helper)—'whose function is to support the other characters to survive' (Gabriel 2015: 298)—while the final roles played are *transformational*, addressing the ultimate status of winner (e.g. hero) or loser (e.g. clown).

Kostera (2012) (after Booker 2004) also analysed archetypes dualistically within an organizational landscape, with different approaches in terms of identifying 'chief' archetypes. However, the basic principle of duality is evident, with each archetype containing two sides, 'a positive imagine on how that role should be carried out selflessly in accordance with instinct, but also its negative aspect' (Booker 2004: 554). For example, the archetype of *self* is seen as the ultimate expression of wholeness, connecting to the selflessness of core instincts. Meanwhile, the *ego* is an entity within the *self*, acting as the hub of consciousness and representing free will 'through which we perceive the world and our own part in it' (Booker 2004: 707). The *self* is therefore 'light', a positive force, the hero, with *ego* being affiliated more with 'dark', with an egocentric archetype distorting perception and limiting understanding. In fairytales, this represents at its most basic the battle between good and evil.

As noted, fairytales are often instigated by 'lack' (see Zipes 1983), representing the imbalance of the *self*, and the quest to restore it. As each archetype has a light and dark side, they represent multiple roles and drivers utilized across a range of contexts and storied spaces. The protagonist's ability to overcome the supernatural opponent is a precondition to reinstate harmony, in recognition that their fates are intertwined. The *self* recognizes the balance in these forces, with the 'fairytale movement from victimisation to retaliation possess[ing] a classic balance and symmetry' (Tatar 1987: 182). In contrast, the *ego* would indicate a villainous role, which uses 'words to intentionally exploit, control, transfix, incarnate and destroy for their benefit' (Zipes 2007: 6). In this manner, the ego aligns to the Jungian archetype of the *shadow*; existing in the collective unconscious, it is often perceived in terms of the darker elements which 'organizations wish to deny about themselves, due to the threat posed to self-image and self-understanding and, more generally, the need to be viewed in a favourable light by others' (Bowles 1991:

378). It manifests, for example, through corruption, misunderstandings, and resignations (Gabriel 2012). However, it is accepted as a part of human life that needs to be recognized in order to facilitate deeper levels of learning and growth. Acknowledging the *shadow* brings the awareness that no organization or employee is perfect: 'the shadow is more intense the more that management tries to gain absolute control over the organization' (Kostera 2012: 65). Supporting Moxnes' (1999b) analysis of deep roles, this can be seen in the *tragic* transformational role, where the loser would always be the winner's *shadow*.

Gender also has an archetypal foundation, with the animus as the unconscious masculine side in women and the anima as the unconscious feminine side of man. The animus displays masculine principles of control, order, and reason. This supports the hero role, which 'personifies those values and epitomizes the strength of the organization' (Jones 1996: 58). It is that ability to take control in times of uncertainty and make sense of complex situations that makes this archetypical image appealing (Kociatkiewicz and Kostera, 2016). It is also arguably an archetype required when business leaders 'seek heroic adventures that demand even more creative courage, imagination, vision and ethical discipline than was practiced by their counterparts in the twentieth century' (Hatch et al. 2005: i).

The dark side, however, is displayed when heroes 'start to believe in their ideal façade if they take their superhuman status seriously' (Kociatkiewicz and Kostera 2012: 867). It can be illustrated by the dual roles associated with an organizational hero, contrasting the disregard for convention of the daring *entrepreneur* (an image stemming from traditional mythologies) with the *company man*, who is reliable but more compliant to formal controls (Gustafsson 1984, as cited in Bowles 1989). The anima, in contrast, 'Jung's term to define the feminine component in a man's psyche which shapes his response to the opposite sex and his relationship to feminine values' (Booker 2004: 708) is arguably associated to female traits of passion, spontaneity, and caring (Kostera 2012). Höpfl (2000) and Kostera (2008a) argue, for example, that organizational concerns with growth strategies and performance metrics, such as (daring) 'continuous improvement' and (caring) 'benchmarking'—operational philosophies and control systems often deployed in FM work—in many ways reflect an imbalance between these archetypes, for 'the constant preoccupation with the organizational masculine aspect leads to the lack of and desire for its femininity' (Kostera 2008a: 166).

These archetypes are present throughout fairytales, and often synonymous with the happy-ever-after motif that is instigated by a marriage, illustrated in tales such as *Cinderella* who weds her prince. This has led to many of the criticisms of the genre—for although male figures agonize, female suffering tends to be greater yet subordinate in stature (Tatar 1987). However, an integration with the *self*, when masculine and feminine values (anima and animus) are in balance, results in the archetypal happy ending—what the heroine needs to make her complete. The anima figure reflects whatever is needed to 'complement the particular balance of qualities shown by the hero. So long as he is worthy of her and open to her, she stands for whatever he needs to make him whole' (Booker 2004: 300).

The good queen/mother remains a background figure in many fairytales and is a symbol of motherhood and loving security (Booker 2004), such as in *Little Red Riding Hood*. Many fairytales 'split the mother image to create two wholly separate entities; a superhuman and diabolical stepmother who is fiendishly spry and vigorous, and a benevolent biological mother who is simply no longer around' (Tatar 1987: 79). In line with the *light/dark father* archetype, the *good mother* protects, the dark attacks or neglects. This binary alignment of good and evil is central to Moxnes's (1999b: 1432) analysis and enables symbolic dynamics to be revealed with 'positive attention and acceptance on one hand, neglect and rejection on the other'.

Similarly, the archetype of the *king* symbolizes authority, legitimate power, and leadership that establishes and maintains order from a masculine viewpoint, with feminine balance displayed in kindness and security as a protective figure (Kostera 2012). Kociatkiewicz and Kostera (2012) further built on the *king* archetype, using it to analyse sustainable leadership. Whereas a good *king* is selfless in their rule, and a source of hope and freedom, the negative represents a heartless tyrant often manifesting as irresponsible and reflecting repression and failure. The dualism is represented in the transformation of the good to the bad, or vice versa. 'Heroes turn into villains if they start to believe in their ideal façade, if they take their superhuman status seriously' (Kociatkiewicz and Kostera 2012: 867). In kindred work, Gabriel (1997) identified follower perceptions of leaders as having both primal *father* and *mother* archetypes, with the former being a figure of authority, reverence, and fear, and the later as caring, supportive, and recognizing.

The *persona* archetype is the image presented to the world: it applies not only to the identity of organizations but also how this affects the identities of the people who populate them (Kostera 2012). It is reflected in attitudes, mannerisms, forms of speech, and types of appearance sculpted for the outside world that are selected for different audiences and situations. It can be detrimental when it becomes rigid and inflexible or is not a true image but one that is felt imposed, for example, manifesting in emotional labour (Hochschild 1983) and thus a sense of disenchantment. Within FM, this can be related to the character of 'veneer FM' (Varcoe and Hinks 2014a).

The *sage* is also identified as an archetype, illustrated as 'an older person, rich in experience who shares their wisdom with the younger generation' (Kostera 2012: 123), which in FM links to the role of consultants. The *sage* represents a 'guide' in that it represents the spreading of knowledge but is a character who does not enforce decisions or actions derived from this knowledge. Within fairytales, this forms the role of the *helper* as illustrated in the actantial model of Figure 3.2.

The *adventurer* archetype is a risk-taker who willingly seeks out new challenges. In fairytales, this is reflected in the protagonist leaving the safe environment of home, facing a series of difficult tasks, but gaining experience and therefore often status. As such, the *adventurer* is associated with transformation (Kostera 2012) and supports the liminality stages as illustrated in Figure 2.2. This new knowledge and experience allow a new status to be endowed and aggregation to be completed. It is acknowledged that this can be perceived as risky behaviour, although the ability to change is also seen as fundamental for societies to function (Kostera 2012). The dark side of this archetype manifests in troublemakers. This has close parallels to the archetype of the *trickster*, which is characterized in the role of perceptual forces, for example, the ability to see and therefore create new ways of viewing reality, breaking accepted norms, and often outwitting authority with cleverness (Raspa 1984, as cited in Bowles 1989). This can be positive, as in the form of a supernatural helper, or negative, as in a cunning, malevolent spirit—representing a liberator or a bandit respectively.

The final archetype is that of the *eternal child*. This archetype is a symbol of hope and rebirth, the capacity for regeneration and renewal (Kostera 2012), with the *eternal child* being the embodiment of playfulness, potential, and serendipity. As such, it is indicative of the more naïve and simple characters in fairytales; those who are 'able to succeed because they are untainted and can recognize wondrous signs' (Zipes 2007: 6). The darker side of this archetype manifests in the inability to change, signally to grow up and accept responsibility and accountability.

Therefore 'deep roles' (Moxnes and Moxnes 2016) allow a clear binary element to be established, while Jungian archetypes facilitate further insight into recurrent motifs (Kostera 2012). In this way, such archetypes can help reveal 'strong plots' (Czarniawska and Rhodes 2006) in that they have been institutionalized and belong to a collective narrative retrieved from common memory. 'We become what we think we are: if the users are repeatedly, and not always verbally, put in the position of the fool . . . eventually they will become one... however some do not need much encouragement' (Jemielniak 2008: 91).

3.6 Coda

This chapter has introduced the concept of storytelling, analysing the role of fiction, and positioning this to fairytales as allegories. In so doing, extended metaphors are discussed in relation to the view that 'fiction accomplishes the feat which organization theory often misses; it combines the subjective with the objective, the fate of the individuals with that of institutions, the micro events with the macros systems' (Czarniawska and Rhodes 2006: 197). The specific introduction of folklore focuses the argument on narrative authenticity—identifying traditional structures to support the credibility of (social and organizational) explanation. An overview of works by Perrault (1697) and the Grimm Brothers (1812–1857)—as 'prominent landmarks' (Warner 2014b: xiii) for the genre—shows how the surreal juxtapositions found in wonder tales can serve to question the plausibility of accounts, in this manner with fairytales representing 'a means of transforming what one sees, of recreating reality' (Holbek 1987: 601).

It is argued that the use of fairytales as a form of organizational storytelling is more robust than the use of standalone metaphors as a research technique, an assumption that is not without academic grounding in organizational analysis (see Baruch 2009) and is illustrated in Appendix 3. Organizational storytelling carries with it and makes effective use of the identification of characters 'such as hero, princess, villain, fool and victim' (Hyde 2008: 49). Research by Moxnes and Moxnes (2016) on mythological roles and attributional processes in leadership development used fairytales overtly as a means of advancing their explanations, while Smith and Simmons (1983: 390) concluded similarly that 'the special contribution of the fairytale to our understanding' is that it represents 'emotional dynamics in a more complex, holistic way than any other system of metaphors'. Such sources variously argue therefore that fairytales can be used effectively to emphasize, define, and promote, for example:

- Magical/fantastical elements:
  Illustrating links to enchantment (see Chapter 2), and allowing features such as absurdity, irrationality, and serendipity to be recognized and included.

- Roles, story arcs, and the BME format (see Appendices 1 and 2):
  Clarifying and explaining complex situations, issues that may be taken for granted, or unconscious behaviours/expectations.
- Clear and identifiable characters:
  Allowing perceptions of roles to be illustrated and the presence of archetypes to be identified, and yet also being flexible enough to facilitate multiple 'deep roles' to be interpreted.
- Universal and memorable imagery:
  Providing familiar yet often emotional similes to make individual, disciplinal, and organizational dynamics/identities explicit, also uncovering behaviours that reinforce wider institutional forces.
- Significance of 'lack':
  Facilitating links to liminality (Chapter 2), with fairytales being used to stabilize or challenge logics, as the protagonist goes through different stages of a journey prompted by the absence of good and/or the desire to have something.
- Recurring motifs and plots:
  Reflecting shared traditions and communal pasts and futures in a flexible and recognizable format—one that allows constant adaptation within increasingly blurry organizational boundaries.
- Extended metaphors/allegories:
  Allowing poetic elaboration so as to incorporate sensitive and emotive areas and perceptions with a view to creating an authentic and engaging tale; also providing a dialogue that may evade censorship.

In analysing the state of FM by means of folklore, certain organizational and conceptual parallels have emerged. Chapter 2 identified themes of a struggle for identity and therefore legitimacy of the practice; a form of analysis in line with many secondary professions, such as HR and IT, but arguably one more salient in the case of FM. This is a line of argument reminiscent of Gabriel's (2000: 130) comment that in many organizational sectors 'contemporary identities are fragmented, tentative, experimental and ever-changing. They are cobbled together in numerous fields'. Storytelling illustrates how the approach helps to explain organizational realities as a sense-giving tool (Collins 2013), with issues such as ambiguity and paradox being incorporated through semi-fiction which allows perception and relationship management to be freely explored. Third spaces, knowledge works, and the concept of liminality as applied to FM illustrate an affinity to fairytale concepts such as 'other', transformation, and metamorphosis. The final section of Chapter 2 addressed the quest for meaning within FM, with process issues, outsourcing, and client expectations sharing characteristics of enchantment, magic, plot, and the role of the audience.

Finally, archetypes and deep roles provide a guide with which to navigate the poetic constructions within a liminal practice and capture fleeting identities that can emerge, develop, and change in secondary industries. The BME format, as well as the familiar and simple structuring of fairytales, provides a suitable frame in which to 'help reduce confusion and ambiguity, and to make contradictions manageable, allowing us to make sense of experience as well as being a means of sharing that experience' (Morrell and Tuck 2014: 2).

4
The Process Bridge

> Man is least himself when he talks in his own person. Give him a mask and he will tell you the truth.
>
> **Oscar Wilde: 'The Critic as an Artist', 1891**

The focus of this study is to investigate FM as an example of a secondary service—a service which has a mutable and often misconstrued identity within the organizational 'kingdom' (McCarroll 2017). The literature review of FM (Chapter 2) indicated a 'shift away from the traditional hierarchies towards more open kinds of network organizations in the economy' (Larsen et al. 2011: 81), this move affecting how FM is perceived and delivered as a supporting industry. In this study, we suggest a new qualitative research approach is needed to understand the role of such support services, and specifically for how the practice of FM can be comprehended by looking at the 'rich symbolic landscape of phenomena such as culture, hierarchy and identity' (Kallio et al. 2015: 391).

This chapter therefore details the new methodological approach we have in mind – *artificial folklore* (as discussed earlier in relation to research aim three), a method which reflects sentiments expressed in the quote from Oscar Wilde above. In detailing this methodology four stages of development are identified: the initial stage is operationalized through creating an in-person research workshop focusing on the deconstruction of fairytales—in this case a session using the tale of *Rumpelstiltskin* as the basis for narrative analysis. Secondly, and building on the in-person exercise, an online collaborative workshop is developed in which the participants are free to build a collective fairytale of their choice. The third phase then sees individual stories created and collected via a two-part interview process. The final stage—of ghostwriting—is developed as a way of illustrating the nature of artificial folklore as formed through the different phases of the methodology.

As noted earlier, the use of fiction is not new within organizational research (see, e.g. Gabriel 2003, Czarniawska 2004, McCabe 2013), although the specific use of folklore remains rare and, in our view, underutilized. This

Once Upon a Time in Facilities Management. Patricia McCarroll and John Hassard, Oxford University Press.
 DOI: 10.1093/9780191943492.003.0004

study therefore builds on previous research using fairytales in organization studies (see Harris et al. 2013, Morrell and Tuck, 2014, Moxnes and Moxnes 2016) in order to establish a method for both unlocking generic narrative patterns within the FM storied space (aim one) and uncovering motifs and archetypes specific to FM (aim two), the overall objective being to form an interpretive picture of the operations of a secondary service within UK industry. What is unique in this inquiry is the use of fairytales, not only as an abstract epistemological philosophy, but also as a means of concrete data analysis, which together come to form the framework for *artificial folklore* research (aim three). This chapter begins with a brief overview of the research framework and its justification before describing in detail the different stages of the methodology (see Table 4.1).

4.1 Overview of the Investigative Framework

There is no single preferred method in the social sciences when it comes to utilizing and analysing narratives in research. Indeed, as sources of inspiration Czarniawska (1998) recommends that several methodological approaches be considered and built on when devising new approaches to fieldwork. In the case of this research, a recognized folkloric framework (after Greimas 1966, Zipes 2011), using literary approaches (after Booker 2004, Kress 1993), has been developed in line with organizational storytelling techniques (after Gabriel 2004, Czarniawska, 1999, Kostera 2006) to form the *artificial folklore* approach.

With the understanding that 'organizations are par excellence jungles of information', and that in this regard 'stories come to the rescue of meaning' (Gabriel 1998: 97), the approach developed here accommodates the role of interpretation very much characteristic of folklore (Benson 2003). This is deployed to help support and understand our previous investigation into the secondary service of FM, and focally the wider role it holds as a support function (Coenen et al. 2010, Price 2011). As designed, the *artificial folklore* framework facilitates a novel research perspective to be created—one that mitigates the possibility of 'dominant reactions' and 'stock responses' (Knowles and Cole 2008) whilst accommodating 'the licence to address potentially embarrassing, dangerous or taboo topics' (Gabriel and Connell 2010: 507). Overall, the aim of the research is the formation of meaningful stories; that is, of narrative explanation in the form of the fairytale—an approach aimed at capturing the warmth and richness of organizational experience rather than providing merely a clinical record of organizational facts (Benjamin 1973).

Additionally, a philosophy of methodological pluralism (Hassard 1991, 1993), or multi-methodology (Goulding et al. 2018), forms the epistemological basis of the *artificial folklore* framework. This is reflected in research that explains, for example, how practitioners can deconstruct an existing fairytale and apply it to various organizational roles they play; how in the same vein they can construct an explanatory organizational fairytale of their choosing at an intersubjective collective level; and finally, how they can create an explanatory organizational fairytale at a subjective individual level. These interpretive objectives have been achieved variously through workshops, online exercises, and face-to-face interviews. Aligned in explanatory terms to understanding the rich narrative variety of folktales, the method reflects conceptually 'the freedom given to each narrator to alter the functions and tasks within the fixed schema' (Zipes 1983: 5).

Returning to our empirical focus on the work of FM, three main service streams were identified for analysis in Chapter 2: consultant, service provider, and in-house manager. These distinctions will be deployed throughout the various phases of the research. These functional approaches are illustrated schematically in Table 4.1 supported by further systematic analysis in Appendix 3 and unpacked in narrative detail as this chapter unfolds.

It has previously been described how the Grimm Brothers (1785–1863) shaped the stories they collected (Chapter 3), capturing not only the tales

Table 4.1 Data collection phases

Phase	Method	Process
1	Deconstruction workshop	One tale—*Rumpelstilskin*—is deconstructured by participants using individual worksheets, but within a face-to-face workshop to assess reaction and answer questions
2	Colloborative online exercise	A story-spine format guides particpants through the creation of a fairytale. This is an online exercise, to increase the potential participant base and allow a wide range of fairytales to be utilized
3	Individual ('lived' and 're-imagined') interviews	A series of individual interviews captures not only 'lived-experience' tales of FM but also allows for a more dialogic 're-imagining' by reference to chosen fairytales
4	Ghostwriting	Results from the phases outlined above are distilled into a single fairytale that represents a particular FM service stream

Source: Authors

themselves, but also what they meant at that time. The tales analysed a nation, whereas this study seeks to scale down this approach to an organizational practice. Essentially what the Grimm Brothers did in nineteenth-century Germany, this study aims to address with twenty-first-century (UK) FM. 'Organizations do possess a living folklore . . . its vitality, breadth and character can give us valuable insights into the nature of organizations, the power relations within them, and the experiences of their members' (Gabriel 2000: 22).

As such, all the research participants represent FM practitioners, who are positioned as storytellers with the understanding that 'when narrators "create" stories, they reveal relationships and build up meaning, as much for themselves as for those listening' (Petit et al. 2011: 395). The fairytale element represents a form of 'narrative truth' (Hatch and Cunliffe 2006), where the validity of the story is established by the connection of the teller to the tale being told, allowing the teller to express things that may not otherwise be acceptable: 'poetic licence is the prerogative of storytelling' a process whereby 'we gain access to deeper organizational realities' (Gabriel and Griffiths 2004: 114). In research terms, this 'science-in-fiction' (Djerassi 1998) approach aims to capture the world of a secondary service by allowing exploration outside what is normally 'prescribed' and 'ordered' (Zipes 1983) in organization science.

4.2 Research Approach

This section details the role of researcher, with recognition that 'no matter how we stage the text, we—the authors—are doing the staging. As we speak about the people we study, we also speak for them. As we inscribe their lives, we bestow meaning and promulgate values' (Richardson 1992: 131). As illustrated in Chapter 1, the first author—who conducted the empirical work—has experience of working both as a facilities manager across a range of industrial sectors and as a lecturer in the one UK university business school offering FM as a subject specialism. This background is perceived as bolstering the research in important ways, not only in the researcher being professionally a 'fellow traveller' to the participants (Kvale 1996) in terms of industrial background, but also in understanding the discipline of FM from a range of practical and academic viewpoints. Given the intimate professional knowledge of the first author to the discipline of FM it is held that the 'researcher–practitioner divide' (Vo and Kelemen 2017) is possibly not as large as in many applied social science investigations. It can be argued that this background feature also supports a folkloric approach in that 'folklore fieldwork is intrinsically a collaborative process between the researcher and participant' (Bronner 2017: 85, see also Jones et al. 1988).

In terms of the wider sociological profile of the research approach, the work of Weber (1947) is influential in supporting the idea that truth can become manufactured—manifesting itself in terms of rationalization and becoming visible in the large bureaucracy (see Section 2.3). This is seen as an important part of organizational existence, but it subsists only as an element in a wider rhetoric, which includes emergent and irrational actions that together form a collection of organizational micro-stories (McCabe 2016). In this research, the related role of 'semi-fiction' (Trubshaw 2002, 2003) is utilized in the belief that truth is a source of deliberation rather than rational thought. 'Narrative truths do not claim to be absolute or eternal, but they do have an extraordinary ability to generate both enlightenment and emotion' (Gabriel and Connell 2010: 508).

The aim of the research is innovative, and concentrates on the participant's point of view, with the process taking recourse both to the instigating and collecting of individual fairytales. Such a storytelling approach follows a 'critical management' philosophy (as illustrated in Chapter 3), with recognition of the multi-vocal and potentially paradoxical elements fairytales can contain. 'Folklore is usually regarded as a reflection of the inherent features of a culture and as an element adapting to culture rather than as an element moulding people's behaviour' (Virtanen 1986: 227). In this way, the ability to hold different aspects of 'reality' at once is acknowledged (see Argyris 2008)—an inherently fluid philosophy answering Cairns (2003: 99) call that 'we must accept that human beings do not consistently hold to a single, incontrovertible "truth".'

The strong influence of social construction is evident throughout the research, in line with the belief that people alter identities via social construction, casting themselves into various characters and roles. Language, and therefore lore, is creative in giving form to reality as opposed to being solely literal (Cunliffe et al. 2004). The storytelling method allows participants to create 'another world' (Buckley and Kenney 1995), building a frame in which specific plots, characters, and therefore archetypes are given definition within FM. In this way, 'storytelling is the transformation of analytical thinking—it takes us onto a more imaginative level, it is a fuller expression of our experience and our perspectives' (Collison and Mackenzie 1999: 39).

In this research, the participants were encouraged to define and control the event, creating a virtual and safe identity. As events are socially constructed, this is artificially assembled within a communally accepted boundary. This interpretative stance, however, is also applicable to the researcher. Although *verstehen* is the empathic medium of understanding, it is recognized that the researcher plays an integrated role within it: 'even a successful theory of knowledge will remain, so to speak, the theorist's own view, reflecting his or her position in the world' (Buckley and Kenney 1995: 36); a situation, in this study, which is supported and strengthened by the experience of the practitioner.

It is this realization that pushes the research towards a social constructionist philosophy, one supporting that 'social realities, identities and knowledge are created and maintained in interactions, and are culturally, historically, and linguistically influenced' (Cunliffe 2008: 201). This philosophy has led to the particular methodology developed here—artificial folklore, which captures moments of action within a fairytale form, thus illustrating how narratives are constructed around specific events and archetypes to expose deeper meanings of practice. The creation of an artificial folklore framework therefore allows the project to adopt a research view similar to that advanced by Gabriel (2015: 730): 'As imaginative researchers, we must generally discover our own compromises between the pleasure principle and the reality principle, between adventure and method, between standing on the shoulders of giants and finding ways to exercise our peripheral vision and let ourselves see things other than those right in front of us.'

4.3 Data Collection Method: The Creation of *Artificial Folklore*

In developing a data-collection approach, participants were asked to focus on fairytales as a way of reflecting on their experience and understanding of working within the FM industry. However, as acknowledged in the analysis of Chapter 2, FM contains a range of professional identities in this regard. Nevertheless, it was arguably still possible to establish them as 'folk' (Gabriel 2000)—and thus as possessing generic and unifying characteristics—given that all voluntarily identified themselves, first and foremost, as FM practitioners.

Additionally, the literature review indicated that, as a support service, FM is increasingly reliant in organizational life on various relationships and perceptions, and notably as they support the casting of 'imagined communities' (Dale and Burrell 2010: 37). The current methodology therefore builds in a way to embrace the different socially constructed realities that make up workplace or organizational practices (Berger and Luckman 1966). In this way, artificial folklore offers a methodological 'alternative' (Hassard et al. 2008) in organization theory to dominant positivist and nominalist approaches to data collection and analysis. The use of familiar characters from fairy-tales support both with Moxnes and Moxnes' (2016) use of 'deep roles' and Kostera's (2012) use of 'organizational archetypes' to unlock an understanding of FM's storied spaces. This approach allows for the flexibility of organizational identity—identity which is 'constructed and reconstructed through a dynamic interaction in which a person is "cast" in an identity by others, seeks to project an identity to the outside world and takes on (or enacts) behaviours, symbols and stories of an identity' (Beech 2011: 286).

As developed, therefore, the artificial folklore framework takes account of such flexibility, allowing FM practitioners to cast their roles within a fairytale setting. This offers an opportunity to redefine or reinforce not only their managerial identities, but also the individual stories and characters that comprise them, thus offering a way to explore multiple realities and capture, in a narrative instant, a range of actors and their voices. The fairytale format lends itself well to this endeavour, with the recognition that stories are often fragmentary in nature, as '*no* story can be the *whole* story' (Baskin 2008: 2 italics in original). However, the BME (beginning/middle/end) format that is so familiar and explicit within fairytale formats often serves to counteract this, essentially providing a mechanism to simplify a complex chronicled entity (Zipes 1997). This narrative research method is also a quite literal interpretation of the work of Ellen O'Connor (2002: 38) and her 'storied business typology' approach, with O'Connor describing how in this method 'the researcher identifies plots into which the organization and its members actively write themselves (as primary characters), as well as plots that they are written into (as secondary characters) by competitors, customers . . . and others'. Fairytales specifically, as a short story, are offered here as a succinct tool to capture the plots of the everyday lifeworld of FM. It is an approach that supports the principles of 'narrative collage' as proposed by Kostera (2006) in her development of a collection of fictitious stories created by the interlocutors.

The term 'artificial folklore' then is created to encapsulate this novel research approach. Within Chapter 3, an exploration of organizational storytelling was completed and the concept of fairytales within folklore evaluated. Here, the vocal nature of the telling of fairytales was highlighted in line with the aural role of the audience. Essentially, fairytales come from a history of oral transmission, one whereby tales were changed to meet the perceived demands of the audience through the role of the storytellers who dispersed them. This meant there was no single author to the tales, as 'they were transformed by the narrator *and* the audience in an active manner though improvisation and interchange to produce a version which would relate to the social conditions of the time' (Zipes 1979: 33). In this way, before being fixed in written form, fairytales continued to hold appeal due to the role and meaning of cumulative collective contributions that organically spanned the ages.

Within this research, however, the organic and multi-vocal emphasis—that places importance on the historic role of the audience—is limited, for the researcher and the audience are synonymous. The creation of the fairy stories has been instigated for a purpose, and produced in relation to a specific request, rather than being captured organically and without time constraints. Essentially the research material collected was not 'encountered in . . . natural state' but instead 'presented and performed for the benefit of an outsider' (Gabriel and Griffiths, 2004: 116). In this respect, the research element sunders the stories, creating a division between literary fairytales—created by a single recognized author—and ones imitating the genre in form, style, and function (Ashliman 2004). As such, the term 'artificial folklore' is introduced to place the stories within a new, predominantly research-based, context.

The creation of the term 'artificial folklore' has been used to address two main issues. Firstly, the term 'artificial' clearly acknowledges the additional elements that have been added to instigate the tales. However, the connotation is with the established term 'artificial intelligence' (AI); for the implication is not that the intelligence is questionable but is merely an indication that the source is not natural or organic; the intelligence itself remains authentic. The phrase 'artificial folklore' therefore follows this precedent, as it is proposed that the stories are representative of a longstanding tale and thus the result of an 'active participation of the people who control their own expressions' (Zipes 1979: 14). The use of the fairytale framework is a means to capture stories at a moment in time. The authenticity remains, for the tale has not been pulled from context, or used to function for political or economic ends (Zipes 2005), but instead is deployed as a tool to further social and organizational understanding.

Second, this acknowledgement serves to differentiate the kind of 'folklore' presented in the current inquiry from that generated for example under the genre of 'fakelore' (de Caro 2013), or 'a synthetic product claiming to be authentic oral tradition' but which is actually 'tailored for mass edification' (Dorson 1963: 94). In the present inquiry, there is no aim to rearrange traditional tales purposely; that is, to alter them fundamentally to create new social meanings, such as in 'fractured folklore' (Zipes 2000). Similarly, there is no intention to 'oppose, re-imagine, subvert, invert, deconstruct or satirise elements of fairytales', as in the genus 'anti-tales' (Calvin and McAra 2011: 1).

As noted, artificial folklore also recognizes the 'liquid' (Bauman 2000) nature of the research, not only from a 'process' perspective of challenging abstract structures and boundaries in social research (Rippin 2013, Kociatkiewicz and Kostera 2014) but also from positions which advance the 'fluid' qualities of folklore itself (Gabriel 2000). This is highlighted in arguments that, for example, roles of researcher, narrator, and audience are existentially intertwined in narrative development and progression. Rather than a research limitation, this is highlighted within the methodology—positioned as echoing the nature of oral history, as the receiver of the story becomes the transmitter, reinterpreting and creating the next evolution of the study. This underscores the philosophical metatheory (Section 4.1) of the research, seen as a 'dialogic process where researchers are never neutral in their attempts to write about the lives of other people' (Rhodes 2000: 511); a position which reflects a continual sense of dialogue that includes interaction with previous information presented.

In terms of the practicalities of the data collection method, initially a website detailing the research procedure was built for participant FM practitioners to consult. This generated a link, which was placed on FM professional social media pages, asking for volunteers; with this being supported similarly by email requests to recognized FM companies. A university ethics application was submitted and upheld at this time—an application which detailed the various stages involved in the research. This document also outlined how participants would prepare for research sessions, with preparation designed to be light touch and optional—essentially a means of trying to increase participant willingness to engage with the investigation. The initial website 'call-out' for the investigation detailed the various research stages and options.

Ahead of the first research session participants were asked to consider—if they wished to—how they would develop their own 'story' of FM. In the process, they were made aware of ten fairytales to contemplate in terms of developing and presenting narratives allied to their professional work. In line with the ethos of qualitative research (Silverman 2021), it was the quality of the stories, rather than the amount of material collected, that was deemed important (Gabriel 2000). A research window of four months was created for this aspect of data collection. Permission was also obtained to run a short workshop on fairytale development for practitioners who were attending a university course in FM, with clarification being given to participants that taking part was voluntary and the information generated would be used for research purposes only. By way of a 'snowball' sampling approach (Saunders et al. 2016), participants were asked to forward details of the research to members of their FM networks, with the only selection criteria for those whose participation was sought being that they identified themselves as FM practitioners.

It is recognized that asking a professional to write a fairy story might have been perceived as unusual. This element, however, was utilized in the research rather than being viewed as a disabling factor. As Bærenholdt (2016: 95) suggests, 'experiencing [which is key to how this research understands storytelling] is about taking inspiration from distractions and surprises to make us liven up and change our ways of life'. The research request was phrased in a way to trigger an element of intrigue and surprise within its approach, thus helping to encourage participant engagement. By targeting busy practitioners, and trying to seek subjective insight into their world, there is recognized recourse to elements borrowed from the 'attention economy' (Bainbridge and Finch 2009). This generally refers to issues of communication and knowledge management, but in this context was used to engage the research participants. It recognizes that attention is a scarce resource, and that people are sensitive to factors affecting their professional survival. It aims to minimize distractibility in terms of engagement with familiar symbols and stories, and the curiosity that the research, on the surface, presents. Lastly, it recognizes the power of engaging people, and securing their commitment to the research by presenting a means for them to create their own stories, but in a safe environment and with an element of co-creation.

On receiving replies to the initial call-out, interested participants were contacted and asked to sign a consent form, with confidentially being ensured at all stages. The nature of the analysis placed a focus on the practice of FM generally, rather than on FM within one specifically targeted company. This appreciation of the FM market removed what could be seen as the restrictive influence of discrete factors and forces, such as the role of specific strategic operations, dedicated organizational structures, or bespoke budgetary systems. Consideration of contingent factors in the wider environment is acknowledged in Zipes's (1979) work, which suggests that any analysis of folktales must consider a range of context markers. In Table 4.2 these are identified and matched to issues relevant to the research:

In terms of an overall methodology, developing a framework to capture fairytales as a basis for understanding the nature of a chosen discipline (FM) is not, in principle, that different from other approaches to qualitative data collection in social science. Collecting such data, and interpreting it

Table 4.2 Context markers

Context requires	Mechanism employed to address context
Background of the narrators	Although each participant remained anonymous throughout the research, they were asked to self-identify with one of three labels: service provider, consultant, or in-house manager, as these were the main professional FM roles identified in Chapter 2. This typology was used in both the deconstruction and construction research exercises, although in the latter additional details were requested.
Background of the community	The fluid and evolving nature of FM as an industry/function is detailed within Chapter 2.
External environment	The United Kingdom emerging from the global financial crisis saw many firms downsize, a situation which affected FM significantly. One factor often treated negatively, and associated with a rise in outsourcing, is an image of FM as increasingly a procurement service (see Section 2.3).
Perspective of the collector	The collector and the researcher are dual roles, and this has been discussed in the section on the research approach (4.1), and again in considering the limitations of the inquiry (Section 9.3).
Reason for collection	The 'artificial' nature of this type of folklore is recognized, with the collection of the tales being used for research into the current state of the FM profession.

Source: adapted from Zipes (1979: 36).

sociologically, is merely a means of creating narrative explanations in order to provide a rigorous research storyline (Saunders et al. 2016). Indeed, this point is inferred by Silverman (2011: 78), who suggests that 'when one reflects on how much of qualitative data . . . takes a narrative form, as indeed do research reports themselves, then the analysis of the fairy tale ceases to look like an odd literary pursuit.' The following section outlines the methodological approaches used to collect this particular type of data in greater detail.

4.4 Stage One: Deconstruction

The first method chosen to understand the viability of using fairytales to shed light on the current state of FM in the United Kingdom was a workshop that focused on just one tale. The aim was to establish not only reactions of FM practitioners to a well-known, yet workwise seemingly unrelated, fairytale, but also to establish if there was identification among the practitioners with the characters and storyline represented. As such, this stage represented the initial part of the inductive and interactive approach to data collection and analysis (Saunders et al. 2016), as key insights were generated that influenced subsequently the individual storytelling stage of the research.

Initially, a fairytale needed to be selected, and the tale of *Rumpelstiltskin* was chosen. This decision was partly influenced by Smith and Simmons' (1983) research, in which this extended metaphor appeared organically by participants detailing the turbulent development of a new organization. In their study, the characters of the tale were mapped on to different departments in order to help participants make sense of the ambiguity associated with a complex, emerging identity. In so doing, the tale not only served to help participants describe their evolving work situation, but also provided them with a mechanism for release of the psychological tension they were encountering in a setting of significant organizational conflict.

The basic storyline of *Rumpelstiltskin* (with possible variations) is as follows: A short tale, with four main characters, it concerns a poor miller who lies to the king to get rich, claiming that his daughter can spin straw into gold. The daughter is promptly imprisoned by the king overnight in a tower so that she may complete the task or face execution (or eternal imprisonment). Seeing her anguish, and knowing she is incapable of spinning gold, a trickster—a little imp or demonic man (Rumpelstiltskin)—offers to help by spinning the gold himself in return for gifts, across what become three nights of increased labour. On nights one and two (and before vanishing each night) he spins straw into gold in return, respectively, for the daughter's necklace and ring. On the third night, however, he spins gold in exchange for her future firstborn, which is agreed, and again he vanishes. Subsequently, after the seemingly avaricious king makes the miller's daughter his queen and realizing a reluctance on behalf of the latter to keep her promise, Rumpelstiltskin says she can keep the child if within three days she correctly guesses his name. After the king is out hunting and accidentally overhears Rumpelstiltskin's name (or a servant secretly follows Rumpelstiltskin and discovers his identity), on the third day the queen 'guesses' correctly, at which point Rumpelstiltskin disappears whilst screaming angrily 'the devil told you that!' (Zipes 2014a: 181).

The major protagonist of this tale is the daughter, 'a young woman from the lower classes, who is generally exploited by her father; the king who marries her; and the little demonic man who tries to take advantage of her' (Zipes 2013: 285). She is positioned as the underdog, both in terms of being female and her class status, but she is not, however, completely powerless, having a great talent in terms of spinning. This was of great economic importance before the industrial revolution, as it was often used as a test to assess the suitability of marriage in rural communities, with a good spinner being able to marry well and therefore rise in social status (Zipes 2000). In this way, spinning was used as a tool and metaphor of empowerment—a means to 'entangle a man and to weave the threads and narrative strands of their own lives' (Zipes 2013: 286). It was only after the rise of automated manufacturing that spinning became regarded as 'hard labour', with much of the industry being completed in male-dominated factories (Dean 1965).

It remains that the daughter is only saved by chance in the tale—marrying the king who threatened to kill her—and so a happy ending for this character does not reflect one of complete choice but of possible compromise (Zipes 1995). In the 1819 edition of the Grimm Brothers' collection, this tale was elaborated so that Rumpelstiltskin tears himself in two at the end, emphasizing the supernatural and monstrous nature of his character, and thus casting a more villainous intent as opposed to earlier, blurrier and more comical, representations (Zipes 2013). He therefore remains an ambivalent and intriguing character, being a supernatural helper (as per the ATU classification) but also a blackmailer. As such, the tale reflects the archetype of the trickster—having the ability to flex metaphorically to both the positive and negative depending on audience perception.

The longevity of this fairytale can be seen in its appearance within the first edition of the Grimm Brothers'collection, *Children's and Household Tales*, published in 1812 (although versions of the tale have appeared in print from as early as 1577: see van der Geest 2010). The Grimm Brothers' version was selected by Zipes (2014a) as part of his bi-centenary collection, where all 156 original stories from the 1812 and 1815 editions were translated and published in one volume. This version also appears within the Aarne-Thompson-Uther classification, numbered at 500, and sitting within the 'supernatural helper' sub-category.

The durability and cultural familiarity of the Rumpelstiltskin character is also illustrated in contemporary fairytales—appearing for example in Philip Pullman's (2012) awarding winning volume *Grimm Tales for Young and Old*. Similarly, for television, Rumpelstiltskin appeared as a main character in the seven series of the ABC television series *Once Upon a Time* (2011–2018) and the six seasons of NBC's *Grimm* (2012–2017). And for film, it is notably reflected in DreamWorks' computer-animated comedy, *Shrek* (after Steig 1990), which has spawned five major box office productions to date.

For the current inquiry, this tale was also selected partially on the basis that the term 'Rumpelstiltskin' has evolved as a 'principle'—specifically one within psychology referring to the power of using personal names (see

Brodsky 2013). Similarly, in the field of psychiatry, the Rumpelstiltskin Principle is used to reflect the power of naming a condition, and focally how this releases the ability to treat it (van der Geest 2010). And in law, the power of a name within a courtroom testimony can also reflect power and influence: 'using attorneys' names in a parallel form of response to questions that begin with our names gives us power. Instead of being linguistically submissive, we can at least be equal. Know the names of the attorneys, judge, and other participants' (Brodsky 2013: 251). This power facility is also illustrated within popular culture, as illustrated in the Harry Potter franchise where the central evil character Lord Voldemort is referred to ironically as 'He Who Must Not Be Named'. This emphasizes his power and status as something to be feared (Rowling 1997), which ultimately is proven when saying his name allows central characters to be magically located, at their peril, as the story unfolds (Rowling 2007).

The *Rumpelstiltskin* story itself has many constituent elements that supported its selection. It reflects many of the classic characteristics that comprise recognizable fairytales—such as the emphasis on the 'rule of three' (a writing strategy suggesting things that come in threes are funnier, more satisfying, or more effective than other numbers of things), and magical elements that are not deemed unusual by the characters (such as the central premise of being able to spin gold). Additionally, the ATU sub-classification of 'supernatural helpers' corresponds well to FM within the helper/service provider orientation. The tale is indeed well known across many fields and disciplines, and it is therefore not unreasonable to expect participants to be at least partially familiar with it.

While the story of *Rumpelstiltskin* is complex in the scope it presents for literary interpretation and analysis, it remains short in length, totalling just 607 words. Within its telling there is no clear hero role, and who is seen as 'good' and 'bad' can be subjective, in line with the trickster archetype. For instance, is the miller desperate and doing his best for his daughter, or is he a liar/malevolent social climber? Does the daughter make rash and foolish promises, or is her true character revealed when she becomes queen? Is the king grasping and avaricious or just fulfilling the role demanded of him? And as for the namesake of the story, Rumpelstiltskin, is he toying with the daughter to steal her firstborn, or does he fulfil everything asked of him only to be ultimately deceived? It is notably the roles of the daughter/queen and Rumpelstiltskin that fluctuate most during the course of the tale, and especially in terms of the exercise of power. It can be argued that 'this is a story of the oppression of a young woman, who is not given the freedom to act available to every male fairy-tale hero', while in the end, however, 'she is heroic, as her endurance and her willingness to cooperate do reach a limit' (Nollendorfs 1983: 92). For reasons of reflecting inherent ambiguities and uncertainties, *Rumpelstiltskin* was deemed a good tale to utilize in the current inquiry. It is one allowing scope for different roles to be identified by the practitioners, and for practical activities and events to be mapped with a range of narrative scenarios.

Returning to the methodological process, once the *Rumpelstiltskin* story was identified as a suitable research medium, a worksheet was developed to support the planned workshop. To establish 'context markers' (see Table 4.3) specific to the workshop participants, the initial question presented was simply one to establish which professional form of FM (see Chapter 1) was being represented:

Would you consider yourself an external service provider, internal service provider, or consultant? Please circle.

To establish the background of the community, the next stage involved 'network mapping', where the participants were asked to note down all the divisions and people they deal within their working environments. This was designed not only to provide insight into the range and form of FM practitioners' professional associations, but also to allow participants to reflect on the culture/environment of their workplaces and the 'characters' within them, thus paving the way for the tale to be introduced.

The next stage of the workshop was to allow the participants to read the story whilst also considering an accompanying instruction:

Read the attached story, Rumpelstiltskin *(Grimm Brothers 1812: Zipes 2014). Within it, there are four characters: the Miller, the Daughter, the King, and Rumpelstiltskin. As you read the story, think about which of these characters you would cast as FM. Use your network mapping to help guide you.*

This is the crux of the workshop: for the fictional characters of a story to be deployed by FM practitioners as a means of 'casting' their work experiences into various roles. So that this could be recorded, Table 4.3 supporting the activity was developed.

The final question asked practitioners if they could rewrite the ending, or any part of the tale, to form a better match with their 'FM story'—also indicating what form this would take, such as the introduction of other characters, circumstances, and so on. This provided participants with a wider analytical scope—ensuring the general framework provided in *Rumpelstiltskin* was

Table 4.3 Deconstruction casting

4. Please complete the table below—explaining why you cast the roles:		
ROLE	CASTING	REASON
Miller		
Daughter		
King		
Rumpelstiltskin		

Source: Authors

used to explore their understanding of FM issues, as opposed to the plot and characters dictating what their views should be. The analysis of this first stage of the research is covered within Section 5.1.

4.5 Stage Two: Online Collaboration

The next stage in forming *artificial folklore* targeted the collaborative nature of storytelling, linking back to the importance of the roles of both storytelling and audience (Benjamin 1973). Here, a fairy story of FM is co-created online by the practitioners. This followed Gabriel and Connell's (2010) experiments using a form of Japanese collaborative poetry called Renga, which also influenced García-Rosell's investigating into the perception of the FM-related role of corporate social responsibility (CSR) in 2016. To support its delivery, the stages of a *story spine* (Dietz and Silverman 2014) are implemented (Table 4.5), which is also referred to as a fairytale structure (Rotmann et al. 2015). Complementary to the structured approach derived from Propp (1928/1968), this provides outline guidance for the participants, who are then free to concentrate on the content of their chosen tale.

As stated above, the call for participants was made through social media (adopting a 'snowballing' method: Saunders et al. 2016) and principally via a purpose-built website. Once participant confirmation was received, an ethical consent form and the questionnaire were sent to the participants (Table 4.4)

The updated form then allowed the assignment of numbers to be used throughout the task to identify the (anonymous) participants. Internally, this permitted the researcher to distinguish between the three main roles of service provider, in-house provider, and consultant.

An opening sentence of 'Once upon a time there was' or 'In a time when wishing still did some good' was provided with an invitation to write the opening paragraph. This was accompanied with instructions to provide the

Table 4.4 Participant characteristics questionnaire

	Question
1.	Gender Male Female
2.	Age <26 26–35 36–45 46–55 55+
3.	How long would you say you have worked in FM? <5, 5–10, 11–15, 15+
Considering your experience of FM to date, and not just your current role:	
4.	Would you consider yourself to be a: Service provider In-house provider Consultant?
5.	What sectors do you predominately associate yourself with? Public Private Third
6.	Would you classify your brand of FM to be: Hard FM Soft FM Total FM Other
7.	What best describes your level of responsibility? Board, Director, Executive, Senior Managerial, Managerial, Tactical, Operational

Source: Authors

fairytale context and central character. These were to be done individually and sent to the researcher, who then released them altogether and asked the group to vote on the preferred opening by ranking them in order of preference.

When the introduction was agreed, participant number 1 was asked to write the next section, that began 'And every day'; which helped set the stage for describing the normal activities that will change. This was issued back to the researcher only, to protect anonymity. This section was then added to the introduction, and shared with the group, so it was clear to all participants how the story was developing. Participant 2 received additional instructions, writing the next section, that started 'Until one day', which introduced the inciting event, and thus the disruption of normal activities. Again, this was sent to the researcher to preserve anonymity, added to the previous section, and shared across the (closed) group. Participant 3 was given the paragraph starting 'And because of this', which opened the beginning of the change, and the start of the journey. This opening was then continued until all the participants had written a paragraph. Participant number 2 was then asked to write a section starting 'Until finally'. When this was received, a call was issued for everyone to write a conclusion starting 'And ever since that day'—to develop the message that is so often found within fairytales. As with the introduction, this element was voted upon via ranking, with the result being shared across the group.

Table 4.5 Collaborative story stages

The Story Spine	Structure	Function	Participant
Once upon a Time	Beginning/initial situation	The world of the story is introduced, and the main character's routine is established	All write this and vote for the favourite
And Every Day			
Until One Day	The event/ story plot	The main character breaks routine	Participant one
And Because of This	The middle/action start/ start of quest	The consequences of breaking the routine. It is unclear if the main characters will succeed	Participant two
			Participant three
			Participant four
			Participant five
			Participant six
Until One Day	The climax	The main character embarks upon success or failure	Participant two
Until Finally	End	The main character succeeds or fails, and a new routine is established	All write this and vote upon the favourite

Source: adapted from Parkin (2010), Gabriel and Connell (2010), Garcìa-Rosell (2016), Propp (1928/1968).

By individually introducing a dilemma and passing it on to another participant, this method allows a plot to organically develop across the group. Each segment is approximately two hundred words as a guideline, allowing a reasonable amount of detail but still maintaining the 'shortness' of a fairytale. Ultimately, this 'negotiated narrative' (Watson 2000) results in a final tale 'whose diverse meaning and silences can be challenged, contested and critiqued' (Gabriel and Connell 2010: 520).

This technique replaces the dialogue between the teller and the listener/perceived audience with a relationship that is formed between the participants, who 'decide, step by step, which knowledge was to be retained and which messages were to be transmitted' (Petit et al. 2011: 401). The role of the researcher was limited in terms of co-creation but is evident in terms of administrating the judged paragraphs and ensuring all voices are considered. There was no direct involvement of the researcher in the creation of the tale itself, an issue which may be considered a weakness of the individual construction technique.

Once the tale was completed, it was followed by a telephone semi-structured interview. This followed a reflective theme (Dietz and Silverman,

2014), aiming to capture the thoughts of the participants not only on the end story but also on the collaborative process. To guide this, four key questions were designed:

- Tell me about your opinion on the story that was created.
- What do you think the story says about the practice of FM?
- Tell me about the experience of creating the story.
- What do you think you will take away from this experience?

The data produced by this approach, and its later analysis, is discussed in Chapters 5, 6, and 7.

4.6 Stage Three: Individual Interviews

The deconstruction of a fairytale, as demonstrated above, led to an understanding of the nature of the FM community. This provided guidance on the next element of the *artificial folklore* approach, which targeted stories being individually constructed. This framework was developed from both organizational storytelling (Czarniawska 1998, 2004, Gabriel 2004, Kostera 2008, Rhodes 2000, Booker 2004,) and folkloric sources (Propp 1968, Greimas 1966, Zipes 2011, Dundes 1980), and counteracted many of the issues encountered in the deconstruction approach. The framework offered enough freedom for the participants to tell their story clearly without it appearing as being too leading in terms of the research objectives. It also provided an academic robustness that recognized the level of interpretation that this innovative and qualitative folkloric approach entails (Trubshaw 2002). As such, the construction framework that was developed contains several stages—which were explained to the participants in order to obtain informed consent and facilitate any preparation an individual felt they needed to complete (Appendix 3). Each session lasted approximately one hour—all were recorded and later transcribed to aid analysis.

To start the session, a quick introduction was completed, re-confirming that all the data collected was to be used for research purposes only, and that it was perception rather than right or wrong answers that were the driver. Due to the researcher/audience roles, a brief overview of the researcher's own story was related. This was a useful icebreaker, but also established a shared history and helped develop rapport (Dundon and Ryan 2009), in that the researcher also held FM practitioner experience.

The initial stage of the session was the completion of a basic questionnaire, which was designed to capture individual details and initiate the start of the research session. As this construction method was a more detailed, individualistic approach than the deconstruction technique, these details were captured to reflect the background of the narrators. This is a specific element discussed by Zipes (1979) in terms of establishing context and is illustrated earlier through Table 4.5. This element was a brief part of the process, as it was mainly collected in case it helped shed light on possible story variations during the analysis section, helping to reinforce the innovative element of the study.

To produce a solid framework, elements of biographic-narrative interpretative methods (BNIM) have been utilized (Wengraf 2001). Within this two-pronged tactic, the participant tells an uninterrupted story of 'lived experience', which is considered as the official story. This is then followed by a second stage, where the participant is asked to create a fairy story that reflects their experience using a developed framework. 'This specific intertwining of imagined and real gives the fictional component of the narrative a sense of plausibility and the factual component a personally approachable, yet "magical" aura' (Shearer 2004: 826). The structure behind this approach supports the idea that the second story sometimes contradicts the official story 'certainly always qualifying it and making the material much richer' (Wengraf 2001: 5). As such, the research as presented here focuses on this second story, the re-imagining of FM, which is designed to capture the perceptions and implicit knowledge (Klein et al. 2007).

4.6.1 Lived-experience Story

As stated, the experiential story was not the focus of the research but was inbuilt to allow each participant to position their FM story in a more familiar format and therefore allow a better connection to the re-imagining stage that was the main element. As such, the participants were asked to relay a story that reflected their experience working in FM, with the initial briefing documents stating this to allow each participant to consider what they wished to say.

A series of prompting questions were also on hand to help the participant recall a story with the recognition that some participants may have found the open question of 'tell me a story' rather 'limiting as opposed to liberating'

(Czarniawska 2001: 42). These were developed from a framework derived from Gabriel's (2000) work:

- Can you recall an incident, which was widely discussed between yourself and your colleagues?
- Can you recall an incident that made you laugh/concerned/sad/proud etc.?
- Are there any special characters in your organization/practice of FM?
- Are there any parts of the buildings or other locations you associate with specific incidents?
- Can you think of an incident that sums up what it means to be part of this practice?
- Can you think of an incident that you discussed outside the workplace, with your partner or friends?
- Can you tell me about your first role in FM/compared to current role?

This approach was underpinned with reference to the critical incident technique (CIT), where participants select an example they wish to discuss, thereby revealing which areas and issues are important to them, through being memorable, meaningful, and so on (Ramseook-Munhurrun 2016).

During this section, the researcher remained quiet; noting down any questions that required clarification, but this was only completed after the participant told their story in full. To help facilitate the next section, notes were also made to highlight specific narrative incidents and to help pinpoint a fairytale typology that was beginning to emerge. As such, specific attention was not only paid to a plot line, but also to the setting, characters, symbols, and overarching themes, in line with the basic plot typologies demonstrated in Appendix 1.

4.6.2 The Fairytale Re-imagining

This lived-experience story then provided the foundation for the fairytale element to be introduced to the participant. As this was viewed as the most unconventional element, it was important that it was presented when the participants were more relaxed with the overall interview process (Butler et al. 2014). Where the factual story was based on a monologue approach, this section was more interactive, but care was taken to position the researcher into a role akin to that of the audience, as opposed to being involved within the fairytale story itself. As such, the ownership of the tale remained with the participant, who was the storyteller, allowing them to freely create their own perceived narrative truths (Enninga and van der Lugt 2016).

The folkloric framework consisted of both a narrative and pictorial element (Stiles 2004), to support the habitually visual nature of fairytales. As seen in Table 3.1 fairytales represent a wide genre, and are constantly evolving and so a reference point was required to identify the most appropriate tales. To establish this, a desktop analysis followed by a workshop with practitioners was completed. The desktop review followed a chronological approach, where the most well-known collectors and then literary authors of fairytales were listed alongside their associated fairytales. To aid identification of fairytales, as opposed to the wide field of folk narratives, the ATU classification system was utilized (Table 3.1), with all the tales falling under the fairytale/magical classifications numbering 300–749.

Appendix 2 details a screenshot of this process, starting with the four classical fairytale writers and their stories: Charles Perrault (1697), the Grimm Brothers (1812–1864), Hans Christian Anderson (1820–1875), and Andrew Lang (1844–1912). This provided the initial grounding of the oral tradition becoming literary, but also spans a good deal of time and geography. To account for the visual element of fairy stories, the Ladybird Well-loved Tales (1964–1990) were also tracked. Modern literary writers were then added to provide balance and contemporary context. To support the mainly UK-based participant pool, UK writers were chosen to help ensure familiarity, although they are all leading authors of their time and represented a wide variety of approaches and re-imaging of fairytales. These were namely, Naomi Mitchison (1936), Angela Carter (1979) Roald Dahl (1982, 1989) Neil Gaiman (1999, 2014), Philip Pullman (2012), Terry Pratchett (1991) and Michael Faber (2005).[1]

The list was then cross-referenced to fairytales that have been mentioned in any FM literature assessment—academic sources and online corporate sources. The desktop analysis was then concluded by ranking the fairytales to illustrate the most dominant across the different categories. From the forty-one that were originally identified, those with less than seven occurrences across the above criteria were excluded, leaving twenty-eight.

A workshop was then held to validate these findings, and to establish the tales that were recognized within the FM field. To do this, the twenty-eight stories were listed in a datasheet, with the request to rank them in

[1] For further details across these authors, please see *The Oxford Companion to Fairy Tales* (2000) edited by Jack Zipes.

terms of *Very Familiar* to *Not at all Familiar*. Twelve FM practitioners (one consultant, four service providers and seven in-house FM practitioners) agreed to complete the form, during a break session in an FM course they were undertaking.

This ranking allowed a further round of cross-referencing, with a final 'top ten' of familiar fairytales emerging.[2] The individual characters from each tale were then identified. This was used as a visual aid to allow participants to connect to the narratives, engaging them with the research, but also allowing a deeper connection by enhancing empathic understanding and generalizability by connecting directly to their embodied knowledge (Goulding et al. 2017). The images were selected from the Ladybird Well-loved Tales Collection (see Sotuthgate 1964–1990), which was the prominent version of the fairytales used. This collection consists of twenty-seven tales, all retold by Vera Southgate, but with different illustrators across the development of the series. Some of the original tales from this series have gone into republication to mark the fiftieth anniversary. Not only are these the most mainstream, visual images of fairytales within the United Kingdom, but they also complement the main age demographic of FM practitioners within the United Kingdom, that is, between forty and sixty years (BIFM demographic report, 2015).

In terms of recognized research approaches, it has been influenced by the *persona* technique, where 'imaginary characters are created to represent real groups of people, which are understood as abstractions or caricatures of people's perceptions of their own roles, or the roles of others around them' (Elvey et al. 2013). The approach is also influenced by the figurative character image elicitation approach, developed by Butler et al. (2014). The latter approach utilizes images of a person or a face, but is also inclusive of cartoon-style diagrams, and is represented here using specific fairytale characters. This helps to create a level of ambiguity between the 'realness' of the image and a familiar outline which encourages participant sense-making. This introduction of visual materials has witnessed a growth within organization and management studies (Hyde et al. 2013, Hoykinpuro and Ropo 2014), and is an inherent part of the nature of fairytales.

Before the framework was introduced, it was clarified that this was a re-imagining (Carter 1979) of a fairytale, and so the original plot could be varied. Not all the characters needed to be assigned, and others could be added at any point, depending on the requirements of the participant's story. They did not have to represent a department or person but could represent

[2] It was noted that the story of *Rumpelstiltskin*, used in the deconstruction approach, appeared within the fairytale selection workshop, but was listed at position eleven, which meant it was excluded from the framework for the construction element.

Table 4.6 The top 10 fairytales and associated characters

ATU	Sub-category	Fairy story	Other characters
310	Supernatural Opponents	*Rapunzel*	Mother, Father, Witch, Prince
333		*Little Red Riding Hood*	Mother, Granny, Wolf, Father/Woodcutter
327		*Hansel and Gretel*	Father (woodcutter), Stepmother, Witch
328		*Jack and the Beanstalk*	Mother, Butcher, Old Woman/Good Fairy, Ogre, Ogre's Wife
410	Supernatural or enchanted relatives: wife	*Sleeping Beauty*	King, Queen, Frog, Fairies, Bad Fairy, Good Fairy, Old Woman (spinner), Failed Princes, Old Man (storyteller), Lords and Ladies, Cook, Scullery Boy, Maid
425	Supernatural or enchanted relatives: husband	*Beauty and the Beast*	Older Sisters (x2), Father/Merchant, Beast/Prince, Wicked Fairy
510	Supernatural helper	*Cinderella*	Father, Step/Ugly Sisters (x2), King, Prince, Fairy Godmother, Rat/Coachman, Lizards/Footman (x2)
531		*Goldilocks and the Three Bears*	Father Bear, Mother Bear, Baby Bear
545		*Puss in Boots*	Millar's son/Marquis of Carrabas, King, Princess, Ogre, Mowers, Reapers
709	Other stories of the supernatural	*Snow White and the Seven Dwarfs*	Snow White's Mother/the Queen, Stepmother/Queen, King, Seven Dwarfs, Huntsman, Prince

Source: adapted from Ladybird's Well-loved Tales series (1964–1990).

anything allocated to the associated characters (e.g. the pumpkin in *Cinderella*). The ten tales and associated characters are illustrated in Table 4.6, as linked to the ATU fairytale categorizations.

The established cast they identified earlier was used to help the participants assign the main characters of their fairy story. A series of open questions were asked to begin to build the story:

Given your experiences of FM, is there any one character or fairy story that you have an affinity for?
Can you explain why you have chosen that?

Allowing a character or a fairy story to be chosen facilitated freedom for the narrative truth to unfold. The visual element enabled the interviewees to use their own language to describe the story and how it related to their experience of FM, capturing their perceptions, and providing an analytical frame. As the story progressed, several prompt questions were available, such as confirming what role they had allocated FM and who else in their story they would like to cast, and in what character. In this way, the fairytale alignment was preserved, being 'both free and controlled at the same time: free (for the narrator must at every moment choose the continuation of his story) and controlled (for the narrator's only choice, after each option, is between the two discontinuous and contradictory terms of an alternative' (Benson 2003: 38).

This re-imagining of a fairytale allowed the specific storyteller to reference their own interpretation of the tales in recognition that the stories are subject to personal judgements (as demonstrated in the different versions of tales as analysed within Chapter 6). Each story developed orally, with the participant using the visual tools and the audience (researcher) in utilizing the story framework to help develop the story where required. In this way, the folklore element of the artificial folklore was preserved. The storyteller's experience formed the tale, and the role of the audience was highlighted to form an evolving, experiential tale within the moment. This oral tale was later written down, forming a literary tale following the path of the evolution of fairytales.

4.7 Stage Four: Ghostwriting

The final stage in the artificial folklore framework (see Appendix 3) combines all the previous elements of data analysis—from the deconstruction, collaborative, and individual construction approaches to allowing a single fairy story for each of the three different FM subgroups to be created. In this way, fairytales are not only used to facilitate data collection but also, as an analytical tool, employed to detect and (re)construct meaning. 'It would appear . . . that after reading a fairy tale, the reader invests the real world with the constructs of the tale' (Zipes 1982: 312). The reoccurring patterns and motifs that constitute folklore (Bronner 2016) form the character position and plot of each tale, which will then be ghostwritten.

The practice of ghostwriting has been described as the 'practice where a researcher engages with a research participant and, as a result, creates a new text that both tells a story of the participant and implies the involvement of the researcher' (Rhodes 2000: 514). In terms of the title artificial folklore, ghostwriting serves many purposes. The artificial element is represented by the fact that all the tales and repetitive motifs were pre-existing in order to be captured in the designed research. As such, there is an argument that their explicit nature was instigated by the research, and therefore did not occur organically. The ghostwriting technique neutralizes this, complimenting the origins of folklore which identifies no single 'true' author, and arguably has at origin a collective creation that is multi-vocal. 'Any name that becomes attached to the written version of the tale—[such as] Charles Perrault . . . theoretically points to the transcriber of a pre-existent narrative and any named source in a collection is again simply a representative narrator of a pre-existent narrative' (Benson 2003: 19).

These short fairytales then represent the storylines and characters symbolizing FM, capturing a unique insight which represents the collective experiences which are 'communicated and institutionalised through repetition' (Rhodes and Brown 2005b: 174). These collective voices are those of the FM participants, a situation which also highlights the role of the researcher as the ghostwriter, which must be noted. It is posited that social research, particularly qualitative, contain mechanisms through which methodology acts as a collector and editor of stories (Kostera 2006), applying frameworks to make sense of a wide variety of data, often ambiguous data. This approach follows such an established protocol but makes the role of the collector and transcriber more explicit (Zipes 1983), underpinning an ethical position (Rhodes and Brown 2005a) and reflective awareness (Kelemen and Rumens 2012) within the overall research.

This design also recognizes the role of semi-fiction, acknowledging that 'the frontier is still difficult to trace as fictional stories are made from real events (producing an effect of reality), while real event are told, often added with many fictional elements' (Breton 2009: 187). This approach emphasizes this argument as a research strength, allowing the patterns—and archetypes of FM to be revealed. The fairytale then creates the space to reframe the stories of FM, mitigating the role of academic privilege and existing hierarchies within data collection (Kelemen et al. 2017). It also removes the participants from their daily discourses—creating new conversations (Ford 1999), and new understandings of this secondary service.

4.8 Coda

As stated, (Chapter 2), the world of FM involves complex contractual, interpersonal, and temporal relationships. To achieve the aim of understanding FM within the United Kingdom, the framework as presented captures lived, factual stories but with a focus on further reflection with fairytales. This use of fiction (Rhodes 2001) allowed a freedom beyond bounded rationality, permitting a deeper reflection of FM to emerge. Fairytales were chosen as the method of data collection, corresponding to a contemporary revival in modern culture, but also because of their ability to simplify situations while maintaining a rich, thick description (Geertz 1973). 'Its figures are clearly drawn; and details; unless very important, are eliminated. All characters are typical rather than unique . . . [although] the fairy tale's deepest meaning will be different for each person, and different for the same person at various moments in his life' (Bettelheim 1976: 13). This provided a relatively consistent framework across four stages—using the creation of artificial folklore to understand FM by seeing it from the perspective of those who belong to that culture (Buchanan 2010). In short, 'familiar stories are always the best' (Lindahl 1997: 265).

5

Beyond the Woods

Investigations

> I think the witch is possibly a big dark cloud of fear, hatred, resistance to change, something cultural in there. It is more of a manifestation of something . . . their biggest fears, worries, maybe of temptation, a devilish character in the business . . . luring them in.
>
> **(C4)**

> Goldilocks probably doesn't speak Bear.
>
> **(SP1)**

> I am the fairy godmother!
>
> **(IH4)**

This study seeks to unlock the narrative patterns within the FM storied space (aim one) and uncover the metaphorical role of archetypes in FM sense-making (aim two). The data collection process has combined fact-based organizational information with art-based research philosophy through the introduction of a fairytale-based research framework—one operating to form an *artificial folklore* method for organization studies (aim three). By doing so, the approach positions FM practitioners as the storytellers of their practice. This method creates understanding through relating factual recollections to the sense-making arising from the fictional casting of organizational performances (Knowles and Cole 2008).

This chapter details how the three designed data collection processes—deconstruction, collaboration, and the individual interview sessions (as detailed in Chapter 4)—were analysed. A total of thirty-eight FM professionals took part in the exercises that comprise the empirical core of the investigation. In terms of the three FM divisions studied, nineteen were from in-house FM, thirteen from service provider FM, and six from consultant FM. Information from these story-based investigations was analysed to uncover repetitive narrative patterns and recurrent descriptive motifs (Tatar 1987) based on thematic, fairytale-based analysis (Reissman 2008, Saunders

Once Upon a Time in Facilities Management. Patricia McCarroll and John Hassard, Oxford University Press.
 DOI: 10.1093/9780191943492.003.0005

et al. 2016). This approach forms the methodological foundation for creating understanding of the collective identity of FM, the forms of its storied spaces, and the metaphorical archetypes of practice that support the aims of this research.

5.1 Stage One: Deconstruction Analysis

As detailed earlier, the initial workshop asked a range of FM practitioners to consider the tale of *Rumpelstiltskin*, using it to cast different roles within their work environment to the characters in the tale. This fictional element allowed the participants the freedom to use the story to project the realities of their practice in a safe format, but also revealed how they viewed their roles by utilizing the ambiguity of the characters within *Rumpelstiltskin*, as explored in Section 4.3.

The participants were all attending a university course in FM, which specialized in organizational communication. As such, they identified themselves as being FM practitioners, as opposed to being labelled as such by the researcher. The class of ten participants was briefed as to the nature of the research exercise, which was voluntary and held during a break time. Six agreed to participate. In line with the planned approach, names were not requested in favour of occupational divisions. As such, the research group reflects the views of three in-house FM's, two service providers and one FM consultant. To aid analysis, they are coded as DIH1, DIH2, DIH3 for the deconstruction in-house participants, DSP1 and DSP2 for the deconstruction service providers, and DC1 for the deconstruction consultant. All direct quotes from the workshop are declared in italics.

Network mapping was used to establish the working environment of the practitioners and then, as a later framework, to cast the characters of *Rumpelstiltskin*. Despite the three divisions of consultant, service provider, and in-house provider, there was substantial overlap in the network mapping. The full responses are illustrated in Table 5.1, with the substantial areas of overlap highlighted in grey.

The consultant network map identifies a range of both operational and strategic contacts (supported by the analysis illustrated in Figure 2.2), ranging from the cleaning function to finance directors. The variety of functions recognized differs across both hard (e.g. contractors/delivery drivers) and soft (e.g. cleaners, receptionists) FM, and reveals that the role of a consultant in this context is multi-functional with no clear single area of focus.

Table 5.1 Network mapping

DC1	DSP1	DSP2	DIH1	DIH2	DIH3
Administrators	Unions	Space management			Head of FM
Receptionists		Helpdesk			
Contractors		Contractors		Contractors	
Cleaners	Cleaners	Cleaners	Cleaning		Cleaners
Security	Security	Security			
		Catering	Legal		
Delivery drivers		Hospitality			
Residents		Visitors	Staff		Staff
Clients	Buyers Clinical directors Nurses	Senior managers			
Builders					Engineers Painters Joiners
IT	IT director IT project manager		IT		
Marketing			PR Communication team		
Finance	Accounts Finance director Operational director Commercial director				
Suppliers				Suppliers	
Competitors			Competitors		
	HR director	HR	HR Health and safety		HR Health and safety

Source: authors.

The identification of finance and competitors in the mapping is notable, emphasizing the contractual nature of FM. This element is accentuated by the absence of HR, which was present in the other responses.

The external service provider grouping support that of the consultant, but with a less overall strategic element. HR, cleaners, security, and end-users dominate this category, indicating a focus on the relationship nature of FM. The main difference within this category is regarding the roles of finance and IT, which dominated one response yet were absent from the other, again indicating the different focuses of FM at service level provision.

The largest group represented within the workshop was the in-house providers. Overall, the number of contacts declared here is greatly reduced from that of their external counterparts, with one participant only reporting two networks, that of contractors and suppliers. This indicates a closed network with limited scope, and a notable absence of any end-users. End-users here are identified as staff by the other participants and would indicate on the surface a distinction between FM as internal members of staff, and of the rest of the organization that they view as more recognized 'staff' members. The identification of PR, Communications and the Head of FM within the other participants positions FM as having wider connections outside the operational scope that is the main dominating theme presented by the in-house operators.

Front-line operational staff are identified across the board, alongside internal departments such as finance, HR, and IT through most of the responses. The identification of the end-users was inconsistent, revealing multiple definitions. This ranged from being recognized as external stakeholders such as visitors, but also as internal staff, showing that end-users is used as a wider umbrella term. Overall, the network map shows that there is a consistency across FM, but the three delivery mechanisms of the professional consultant, the external service provider, and the in-house function revealed different areas of focus. This would indicate three diverse storied spaces contained across the discipline.

By completing this network mapping, the participants had also simultaneously created their cast for the fictional tale to be introduced. As discussed, the tale of *Rumpelstiltskin* is short and contains only four characters, but these can be interpreted as to their benevolent or malevolent nature in line with the binary nature of archetypes (Kostera 2012, Moxnes and Moxnes 2016). To analyse this section, each re-casting story is taken in turn and analysed with reference to the basic plot typologies (Appendix 1). As the tale of *Rumpelstiltskin* remains constant, the typology was utilized to establish key elements in the retelling, for example, who was positioned as the main protagonist, which representative archetype was adopted, how the plot was perceived. In other words, does this change due to the viewpoint of the chosen protagonists, the emotions that are generated, and the typology and so on. This would indicate how the practitioners view FM within organizations, which was then compared to the factual mapping. Finally, the individual retellings are analysed as a collective.

5.1.1 The Rumpelstiltskin Consultant

Story one, representing a consultant (DC1), cast the two main roles of the daughter and Rumpelstiltskin within FM. The daughter is seen as the FM consultant who is expected to deliver what many consider unachievable, by 'creating something from nothing' (DC1). The element of luck, a dominant feature of fairytales is also referred to, with 'finding the solutions to key issues sometimes by chance' (DC1).

The role that the FM consultants identified with is the *Good Daughter/Princess* (Moxnes and Moxnes 2016). In this deep role, there is a recognized skill that is well respected and sought after, and yet her salvation comes by chance. This also aligns with an imbalance within the animus and anima archetypes, with the animus being largely absent.

Rumpelstiltskin is front-line FM, and the element of invisibility is highlighted by this casting. They 'deliver what is asked of them—sometimes in impossible situations—and they are unnamed' (DC1). This is somewhat paradoxical as the front line is the most physically present FM function, and it is not clear if this relates to in-house or external service provider provision. Rumpelstiltskin is not cast as the villain in this tale, favouring the supernatural helper casting, and indication of the *trickster* archetype. However, the deep role of *spiritual helper/false prophet* shows the darker side of this character: he is supernatural in his invisibility but has great power to achieve all that is asked of him, without any thanks and ultimately to his detriment.

The miller and the king are both allocated the external role of the FM client, having the never-ending power to demand impossible solutions, wanting 'more and more constantly' (DC1). The miller, as the client, has unlimited and legitimate control of FM—as the father figure, who is willing to use his daughter as a commodity without giving much consideration to any ill-effects his actions may have on her. This villainous side to the client is emphasized by also being cast as the king, who again holds a position of legitimate power in his constant and often unrealistic demands.

The plot is therefore focused on the never-ending needs of the client and FM's eternal role of being subservient. The emotions that this casting generates are sorrow, pity, anger, and blame. By comparing this storyline to the network map, where a large range of characters were identified, the focus as perceived by the consultant is the FM delivery arm. This is an invisible yet skilled force at the mercy of only one contact—that of the client. This would position the consultant tale of *Rumpelstiltskin* into the *tragic* categorization (Appendix 1), a role surrounded by dark characters.

5.1.2 The Rumpelstiltskin Service Provider

The first external service provider (DSP1) cast FM as the miller within the tale, with the association that FM, like the miller, is very poor and has therefore been reduced to the current predicament of using his daughter as his last remaining resource. There is no malevolence displayed here, rather more a sense of desperation, as he is forced 'to sell the family silver' (DSP1). The statement to the king about his daughter's ability to spin straw into gold is therefore an example, not of deception, but rather that he is 'hoping for a miracle' (DSP1). However, utilizing deep roles, the actions of this character would reflect the negative father role (Moxnes and Moxnes 2016), thus a placing it as the archetypal *shadow*.

The role of the daughter is assigned to in-house FM—a role liaising with the external service provider in terms of the original contract and ongoing service-level reviews. The in-house FM as the daughter, reveals their initial 'lack of control at first' (DSP1), but as the story evolves, they realize the power that they hold, which was originally underestimated. This indicates a lack of awareness on behalf of the internal client function, but a willingness and ability to learn from the changing landscape—a situation evolving 'to be become very powerful at the end'. (DSP1), moving from being a miller's daughter to the queen. In archetypal terms, this represents a balance of the animus and anima. Along with this perceived shrewdness is the strong recognition of luck—that the help that was received was not planned and is instead a result of providence.

The character of Rumpelstiltskin is assigned to the client, who 'want(s) the treasure' (DSP1). This indicates an element of being unobtainable—the service they want may not necessarily be the one that was bought by the client. It also carries connotations of unreasonableness and inflexibility, but this is counteracted by the fact that 'the rules kept changing' (DSP1), reflecting a fluctuating and confusing environment. This element of powerlessness also indicates a 'lack' in the animus archetype. In the tale of *Rumpelstiltskin*, this is the three exchanges between this character and the daughter and would indicate possible communication issues with the internal client and end-users—exasperated by the initial powerlessness as associated with the client at the beginning of the tale. The lack of control prescribed to the client/in-house FM provider through the casting of the daughter is not clear. With the service provider FM allocated to the poor miller, it is unlikely this issue of power stems from a private service provider hired in as an expert, and exploiting internal systems, but rather this is powerlessness in the face of the internal pressures of finance, as illustrated by the casting of the king.

This last role, of the king, is attributed to the director of finance. This casting is very rich in its terminology, emphasizing his unreasonable nature; demanding, for example, 'blood out of a stone (money)', with endless 'cost saving' reflected in the actions of a 'dictator' (DSP1). This casting is the clearest archetype present in this story. The characters of the miller and Rumpelstiltskin could be argued to be ambivalent, but their actions indicate the *shadow* archetype. It is internal FM, as the daughter, that affiliates most to the hero archetype—facing many obstacles along her journey but being ultimately successful. The king meanwhile is clearly cast as the villain.

This tale had the addition of other characters, displaying how this service provider would seek to alter the story. In this retelling, the daughter/internal client would not be powerless from the start but would have additional help in the form of procurement, which would negotiate on her behalf for better initial terms and provide a further level of understanding. This is a point of interest as although this tale is written from the perspective of the external service provider—who is cast as the downtrodden and desperate miller—it is the in-house client that receives the additional help. It could also be argued that this represents how this practitioner positions FM, as accepting the situation. Another layer of protection is allocated to the daughter/client, in receiving security to stop Rumpelstiltskin's involvement, which is also rendered unnecessary by the empowerment properties of procurement.

In terms of the basic plot typologies (Appendix 1), the external service provider does not see themselves as the main protagonist, who is the client. In-house FM is the daughter who is then positioned as the hero, as the story unfolds. From this viewpoint, the emotions associated are pride and admiration—as the protagonist gains power, casting the tale into the *epic* category. However, this is the service providers' tale, and from their viewpoint, they remain as the destitute miller, with a deep role of a bad father, evoking a certain degree of sorrow and pity that is more associated with the *tragedy* categorization.

This would support the factual network mapping in that the service provider noted many roles associated with the in-house company and few associated with the service provider company itself. It also supports the vision of external FM receiving few resources, symbolized by the miller. The most dominant function is finance, with the account director, finance director, operational director, and commercial director all being mentioned, illustrating their collective power base and dominance within this tale.

Service provider two (DSP2) also positioned the daughter as in-house FM, possibly identifying with the powerless quality displayed by this character in the tale: 'then she was locked in the room where she sat and wept'

(Zipes 2014a: 181). This is emphasized by the casting of the role of external FM as the miller. Unlike the passive role, as displayed in DSP1, the service provider FM are actively 'offering service they cannot provide' (DSP2), which signifies a more malicious and therefore villainous character, one reflecting a dark *trickster* archetype. The king is cast as senior management of the service provider company, with the front-line staff assuming the role of Rumpelstiltskin. The participant only detailed the reason behind the casting of the miller/DSP1 character, making it difficult to analyse as opposed to project, and so further reference to the network mapping was required in this instance. This revealed an emphasis on front-line staff, such as the helpdesk, cleaners, security, and catering, which would indicate the degree of attention this element reflects in the world of DSP2. This would, at a surface level, also reflect the importance of this function—aligning it with the more malicious side of Rumpelstiltskin.

The application of the basic plot typologies would move the main protagonist to the DSP2, with its association with the miller, as this was the only character that was allocated supporting information. This would generate emotions of possible aggression, positioning the DSP2 as a *trickster*, indicating an overall story plot with that of a *comedy*. Where DSP1 saw their role as service providers as powerless in the miller role, with admiration of the in-house teams, here the service provider tale presents themselves as the dark *trickster*, a deceitful entity. However, it is noted that this participant provided little information to support this analysis in comparison to the rest of the tales submitted.

Overall, the external service providers' casting of Rumpelstiltskin presents a picture of this role being either powerless or borderline malicious within their story projections. The hero role is most closely associated with that of the internal FM provider, who is powerless but gains status as the tale/contract unfolds. The villain takes the guide of senior management, pinpointed though the function of the finance department.

5.1.3 The In-house Rumpelstiltskin

This category represents the biggest sector of FM as represented in the workshop. The main protagonist in the tale, the daughter, is not represented by a department in the first retelling of in-house provider 1(DIH1), but rather manifests its characteristics through the building that FM is charged to manage. This imaginative association positions both the daughter in the tale, and the building within FM delivery as the product, and as such, the physical manifestation of FM that demonstrates the skills and charms of service delivery, ultimately allowing 'her to get her happy ending' (DIH1). Rumpelstiltskin is the main character that this in-house storyteller relates to—who is seen 'to offer hope and kept his word, got no recognition and disappears at the end' (DIH1). It is therefore clear that in this casting, Rumpelstiltskin is not the villain but the secret hero of the tale, looking after the daughter/the physical environment. He is transparent in his contract and fulfils his obligations but has no happy ending despite his honesty and work. There is an exclusion of the game-playing nature of this character found in the original tale (Zipes 2014a), therefore signifying the *wiseman* deep role and the *sage* positive archetype.

In similarity to the previous retellings, both the miller and the king are interpreted as more wicked entities, with the miller 'lying to get what he wanted to get a return or to get noticed' (DIH1). As the original tale has the miller exchanging his daughter for the king's favour, this aspect was focused upon, resulting in this storyteller casting the role to the procurement manager. The king who is 'greedy, demanding, powerful and unrealistic—a dictator' (DIH1) is seen as representing the senior management team.

In this tale, there is a clear dichotomy in the delivery arm of FM, the daughter and Rumpelstiltskin who are skilful and honest and that of the internal hierarchy of procurement and the senior management teams—those who are not considered to be part of FM but to whom FM answers. While this storyteller has no happy ending for their role, the product of the service delivery—the building—does speak for itself and stands as the real recognition of the service.

An addition in this story comes in the form of a king's guard, who protects the daughter/the building, which is FM's ultimate purpose: 'so the miller (procurement) would not have been able to sell a false story—they [the king's guard] would act as an audit role to validate any data' (DIH1). Despite procurement and senior management taking prominent roles in the casting of *Rumpelstiltskin*, there is no mention of these departments within the initial network mapping, although it is interesting to note that competitors are highlighted. This category only appeared in one other map—that of the consultant. Uniquely, this network map made reference to communications and PR functions. Together with an association with competitors, this may indicate that the pressure to outsource in-house FM provision may be an underlying theme to this interpretation of the tale, with procurement seen as the villain of the piece and thus assuming the *shadow* archetype.

The main protagonist of this tale is interpreted as Rumpelstiltskin, who is represented by the DIH1, and is cast as a helper that is honest but receives no recognition, to the point of invisibility. As such, the emotions are once again linked to sorrow, pity, and fear, placing this re-imagining into the *tragedy* genre of story typology.

Taking the main protagonist, the daughter role, and analysing the second in-house tale (DIH2), reveals the symbolism to the front-line FM staff. This is specifically identified as the service provider staff, who are seen to have 'little control, receive instruction [and have no] choice' (DIH2), Despite that, they still are able to negotiate a deal that is suitable for them. Rumpelstiltskin is again cast as an external entity, that of the service provider senior management, which provide the front-line staff under contract (the daughter). In this tale, the FM service providers can provide what is requested, but are not passive, expecting to assume a role that allows for negotiation and change.

The two remaining roles of the miller and the king are cast internally, to the senior FM manager and the director respectively, and both are positioned negatively—relating to the *shadow* archetype. The miller 'offers something that isn't deliverable when trying to impress a superior' (DIH2), who is the director/king, who in turn 'looks for the final output but not the process to achieve it' (DIH2). The author of this tale also associated characteristics to the four roles, with the miller/senior FM manager and king/director being declared as 'bad', the daughter/external front-line staff as 'good' and the overall service provider/contractor as 'mixed'.

The participant representing the in-house FM viewpoint chose not to give in-house FM a specific role in the fairytale, indicating a voiceless symbolism. Due to this, the main protagonist of the story remains associated to the original, defaulting to the daughter and therefore the role of front-line external staff. Here, the protagonist sets out to on a quest, 'facing many obstacles along the way' (Booker 2004: 69), as an unwitting *adventurer* archetype. The emotions generated by this protagonist are ones of admiration, placing this tale within the *epic* category (with the sub-category of the *quest*). This casting, as utilized to re-imagine the tale of *Rumpelstiltskin*, provided much richer detail than the network mapping, where only two departments were identified, that of contractors and suppliers. By analysing the story, it would indicate a focus on the external FM sector, supported by the storyteller's perception that this is 'good', whereas internal FM is at the opposite end of the spectrum. However, the fact that the storyteller's direct role in the tale was omitted cannot be ignored, casting a slight question over this categorization.

The final in-house provider (DIH3) cast the daughter as the client/end-user, making a different interpretation of the story. Here, the daughter 'has a task to do but needs FM to be able to complete it' (DIH3). Correspondingly, Rumpelstiltskin is moved to the main protagonist role and takes the form of FM, which 'does all the work in background but gets no reward' (DIH3). This tale links these characters together but places the power within FM as an essential component to the success of the daughter; but this is not given recognition by the client/daughter, who accepts the help but does not acknowledge it. The miller and the king are again cast in the *shadow* archetype, with the miller being marketing and sales, who 'promise everything without checking to see if it can be delivered' (DIH3). The king is senior management, who 'sees something that isn't valuable but can be made valuable at no cost—wants more and more' (DIH3). This indicates an insatiability, as illustrated in the tale, where 'the king found the entire room filled with gold, but because of this, his heart grew greedier' (Zipes 2014a: 181). This tale is more insular looking than the re-imagining that went before it—there is no mention of any external influence, and this is reflected in the network mapping. Once again, the two main characters of marketing/sales and senior management do not appear within the initial factual mapping. This tale positions Rumpelstiltskin as the main protagonist, but the plot is one of 'misrecognition which is often underserved' (Hatch and Cunliffe 2006: 199), generating emotions of sorrow and pity associated to the overall category of *tragedy*.

Taking the in-house FM Rumpelstiltskin as a collective, there is a dominant picture of FM as the character of Rumpelstiltskin, who is portrayed as a good, heroic character in that his intentions are good, and his behaviour is honest—he is clear on the service that he can provide and completes the tasks as promised. This is the *spiritual helper/wiseman* deep role and the *sage* archetype. However, this role holds no recognition, and indeed, is to FM's detriment, leading to an unhappy ending for FM. Evidence of hierarchical power influencing FM's position is seen in the role of the king—cast as senior management across the board, indicating a fundamental misunderstanding of what FM is charged to do, and how it is therefore not regarded as having strategic importance. The idea of getting scant reward or acknowledgement for good work done is strongly apparent. The senior management team that FM answers to was cast unanimously as the king, which was deemed to be a powerful yet greedy character, ignorant in how processes are achieved and representative of the *shadow* archetype. The daughter's role is cast to front-line staff and end-users, who have little power and are reliant on the services of in-house practitioners.

5.1.4 The FM Rumpelstiltskin

The flexibility demonstrated in the ambiguity of the tale of *Rumpelstiltskin* allowed the different storied spaces of FM to be captured. In summary, the full casting is mapped in Table 5.2 with the grey shading highlighting into which role the storyteller cast themselves.

This approach asked FM practitioners to deconstruct the story of *Rumpelstiltskin* to reflect their occupational world. Each story is analysed above to reflect the different areas of the practice they work within. Taken as a whole, each story can be cast back into the main tale itself—one fairy story that can help understand the practice as a whole—containing the different characters and divisions. The in-house facility managers identify with the role of Rumpelstiltskin, interpreting him as a fair character that is clear in what he can deliver, is open to negotiation, but has no recognition and is helpless in the face of senior entities. This does not however make them powerless, illustrated in the dependence of the daughter role on this character, but it does place FM in a deep role of a *helper*, albeit a *wiseman* (Moxnes and Moxnes 2016) and archetypal *sage*.

The miller is the external service providers: over stretching their promises, making the client (daughter) lucky to survive because of their actions. This could be indicative of 'veneer FM' that Varcoe and Hinks (2014a) warned of in Chapter 2, the *shadow* archetype. The motives behind this differ, with one version relaying the desperate situation of the miller who is selling everything he has to survive, 'hoping for miracles' (DSP1). In the other tale, the miller's intentions are more deceitful and aim to exploit in-house FM. Both external service providers' tales illustrate the precarious nature of their practice and how reliant they are professionally on in-house FM for survival. There is also a hint of admiration towards in-house FM, as this function is predominately cast as the hero in the service provider's retelling.

The FM consultant claims the role of the daughter. The unrealistic demands they face are well emphasized, as they navigate their way through the utilizing of knowledge, often relying on luck. They remain anonymous for the most part and remain at the mercy of the miller and the king, who dictate their actions and constantly demand more. The tale of *Rumpelstiltskin*, as cast through FM eyes, creates three storied spaces, identifying a common theme of oppression and endurance (Nollendorfs 1983), but not necessarily of a happy ending.

It is understood that this is a generalization from a small sample, although the academic analysis of Chapter 2 supports many of the empirical findings. Clear patterns such as the issues of impossible tasks with no recognition or

Table 5.2 Deconstruction casting

	DC1	DSP1	DSP2	DIH1	DIH2	DIH3
Miller	**FM client** Puts FM in a difficult position, asking for impossible solutions.	**FMSP** Poor with no money. Selling the family silver. Hoping for a miracle.	**FMSP** Offering services they cannot provide.	**Procurement manager** Lied to get what he wanted—to get a return or get noticed.	**Senior FM manager** Offers something that isn't deliverable when trying to impress superior (bad).	**Marketing/ Sales** Promises everything without checking to see if it can be delivered.
Daughter	**FM consultant** Create something from nothing and deliver the impossible. Finds the key issues sometimes by chance.	**In-house FM** Lack of control at first. Very lucky. Underestimated power.	**In-house FM** Not detailed.	**The building** Seen as a product but uses her charms and skills to get her happy ending.	**FM front-line staff SP** Little control, receives instruction, no choice, negotiates a deal (good).	**Client/User** Has a task to do but needs FM to be able to complete.
King	**Client** Wants more and more constantly.	**Director of finance** Wants 'blood from a stone' (money), and savings. Very powerful. Demands improvements. Dictator.	**Senior mgmt.** Not detailed	**Project board (Senior mgmt.)** Greedy, demanding, dictator: powerful and unrealistic.	**Senior mgmt. (director)** Power: looks at the final output but not the process to achieve it (bad).	**Senior mgmt.** Sees something that isn't valuable but can be made valuable at no cost—wants more and more.

Continued

Table 5.2 *Continued*

	DC1	DSP1	DSP2	DIH1	DIH2	DIH3
Rumpelstiltskin	**Front-line FM staff** Un-named staff who deliver what is asked of them, no matter how impossible.	**Client** Wants the treasure—rules were changing.	**Client** Not detailed	**IHFM** Offered hope and kept his word—no recognition, disappeared at the end.	**FM consultant** Can provide service but demands something in return. Did as instructed, allows for negotiation and change (mixed).	**IHFM** Does all the work in the background but gets no reward.
Rewrite Options	N/A	Third party to negotiate for the (childlike) procurement. Bigger role for security to stop Rumpelstiltskin.	N/A	Inclusion of a king's guard, so the miller would not be able to sell a false story: would act as audit role to validate any data	N/A	N/A
Typology	Tragedy	Tragedy	Tragedy	Tragedy	Epic	Tragedy
Archetype/Role	Good daughter/ princess Anima (lack of animus).	Bad father/Devil *Shadow.*	Bad father/ Devil *Dark. Trick. Shadow.*	Wiseman *Sage.*	Good daughter/ princess *Adventurer* (View of SP).	Wiseman *Sage.*

Source: adapted from Booker (2004)

suitable reward, and the dominant role of finance, are apparent. Despite representing three of the main roles within the story, FM remains the underdog in the majority of tales, a dominant story plot of *tragedy*. FM have limited power, reflecting operational reality, which also would indicate that strategic FM remains a fantasy. Elements of liminality are apparent in all the retellings but most dominant within the in-house FM tales. This may be a dominant theme within this type of FM, but it may also indicate that the consultant and service providers accept this as a condition and therefore this element is not emphasized.

This initial approach, in which one fairytale was pre-selected and then deconstructed and recast by the practitioners, as opposed to constructing a fairytale of their own invention, offered an insight into the feasibility of the approach, as it is easier to work with an existing document as opposed to starting one afresh. This exercise in data collection also proved a useful tool in designing the construction phase of the research. Although the researcher was available to answer any questions, further contact with the participants, at an individual level, would have helped clarify some of the responses, notably to understand more fully the meaning that each participant attached. The end request of providing an alternative ending, or reworked tale, needed more time to be considered, for although this was an effort to move away from the prescription of a preselected tale that practitioners needed to accommodate, it was generally not answered.

Each participant seemed to enjoy the experience, and the later analysis of the worksheets offered a snapshot of the different perceptions of FM as forecasted. This indicated that using folklore, specifically fairytales, represented a useful framework with which to discover the storied spaces of FM (aim one) and their character archetypes (aim two). The combination of factual network mapping and fictional casting worked in synergy to produce a fuller picture than use of metaphors alone would permit. This approach allowed a different way of looking at the storied whole, with the framework providing an insight into the narrative parts on which it was based (Smith and Simmons 1983).

5.2 Stage Two: Collaborate Online Analysis

The collaborative approach aimed to recruit six volunteers to contribute to the research (as illustrated in Appendix 3). This was done by utilizing the website that also detailed the individual storytelling interview option. Its aim was to allocate the power of the storyteller directly into the hands of the participants, as they negotiated the narrative as it evolved through different viewpoints (see Petit et al. 2011). The characters, plot, emotions, and overall typology (Appendix 1) were completely within the participants' control to create individually, but as a continuation of the previous entry. The introduction and conclusion were to be voted upon, emphasizing the freedom of the writers but also the collaborative nature of the end product. This also allowed the rejected and silenced narratives (Gabriel and Connell 2010) to be captured to provide a robust overview of the FM industry.

While the initial call for participants resulted in a favourable response to the individual, interview-based data collection approach, in contrast there seemed less initial interest among potential participants in this more collaborative approach. To bolster this approach, an additional ninety-two personalized invites explaining this research process were emailed, utilizing the individual connections within the professional LinkedIn network.

This resulted in one story introduction being submitted, and this is captured below. Due to the lack of numbers, the full approach as planned was deemed unfeasible, and as such, the approach utilized within the deconstruction data collection was employed. This offered a sense of consistency, but also reflects the planned initial stage of analysing 'Renga' as introduced above (Gabriel and Connell 2010). The lack of response of this approach is then also considered.

The single submission is the introduction (see stage one, Table 4.5) as provided by an in-house provider; a male aged fifty-five plus. Mainly associating with soft FM within the public sector, he held a position at an executive level and had over fifteen years' experience within the FM sector. The story as received is illustrated in Figure 5.1.

The function of the introduction was to introduce the context of the story and establish the main character's routine. Although this is the only section of the story that was received, it is worth noting its main characteristics in a similar manner to the deconstruction analysis. In-house FM is positioned as the main protagonist in the story but is without any 'power' and not in charge of its own storyline, once again illustrating the imbalance of the animus and anima archetypes. Other characters, such as senior management and four other departments are identified, but FM is subservient to all of them. Not only is there active resistance evident, but also a level of confusion as FM is caught in a power battle between the other divisions. Despite this turmoil, FM is cast in the hero archetype, aware of the hopelessness of the situation but still striving to deliver a reliable and ultimately safe environment for all. There is also an element of *comedy* with the imagery of being 'headless', as they run between reactive situations being salient.

> *Once upon a time ... there was a team that worked in a really challenging environment, facing huge budget cuts, wanting to be a corporate landlord but facing resistance from the senior management team, wanting to let cost-effective and efficient contracts but struggling to understand the requirements of the four departments it serves. The head of property is equally perplexed, sitting in another department and not having the strategic overview of activity across the estate, which is causing huge problems as individual departments seek to try to generate income by letting property or space without clear controls.*
>
> *It was a department that wanted to embrace modern technology, bring innovation into the workplace but faced resistance to change from all quarters and despite every effort it continued to deliver a service that was reliable and made every effort to ensure compliance with legislation.*
>
> *The head of FM was called 'headless' — they charged around — dealing with stakeholders, both senior and junior, negotiating contracts, monitoring compliance with business rules, legislation, and standing orders, recruiting replacement staff both agency and permanent, challenging spend and meeting tough saving targets, meeting internal income targets, and driving modernization and change.*
>
> *The fairy house above is where we work, the windmill drives power and lights up when it's dark — but is there light after dark in this world of challenge and inefficient use of buildings, contracts, and manpower?*
>
> *Where can we find an answer?*

Figure 5.1 Constructed in-house tale

Source: In-house FM participant.

The introduction ends by posing a question, utilizing the striking symbolism of light in the dark, positioning FM as being able to literally 'keep the lights on' but not for much longer. Here the emotions generated are those of frustration and determination. There is a drive to succeed but doubt over the likelihood of success. In this way, it supports the tales as told by in-house FM in the deconstruction of *Rumpelstiltskin*. At this early juncture of the story, the plot typology would indicate a *tragedy*, supporting the story arcs as identified within the deconstruction analysis.

It is recognized that the incompleteness of the story leads to a level of conjecture, and so to address this level of incompleteness, the lack of response to this research option is addressed. Gabriel and Connell, writing about their own experience of writing a Renga as an academic revealed they found it very difficult, as it requires 'very different faculties of imagination, creativity and storytelling than what academics do a lot of the time' (2010: 519). Placing this into the context of FM, Chapter 2 analysed the legacy of the practice as emerging from a background of IT and engineering, which lead to a dominating aspect of the 'hard' services focusing on the physical environment. This is seemingly at odds with the strategic element of FM, which is pushing towards more people-centric delivery and has relationship management at its core. These differentiations on what the functions of FM are then contributes to the wider identity issue of the practice.

The lack of response to this approach may indicate that there is still a dominant leaning towards the tangible management of the physical environment as opposed to the intangibility of people management. Supporting the earlier research using this method, the lack of response may reflect there is 'no inherent interest in storytelling and narratives . . . for example, (by) those with strongly quantitative, statistical, and factual leanings, who may not respond warmly to such a prospect' (Gabriel and Connell 2010: 520).

However, it must also be recognized that the collaborative nature of this approach could lead to anxiety among participants as 'sharing stories in public is not very common and can evoke feeling of fear and shame' (Abma 2003: 230). This element may have been exasperated if the prospect of producing a cohesive plot was already daunting, and a tragic storyline, as seen in deconstruction, may not have been perceived as an appropriate tale to share. As full-time practitioners, the reactive nature of the role, as supported in Chapter 2, and then within the deconstruction analysis, may also have impacted on this proposed research, as the longevity of the approach may have been seen as quite time-consuming.

5.3 Stage Three: Individual Interview Analysis

In total, nineteen participants volunteered to take part in this approach. Each session lasted approximately an hour, with permissions granted by all participants to record the sessions for later transcription. As with the previous stages, a practice label was applied to code the participants, therefore preserving anonymity. The letter C was utilized for consultant and then numbered in line with the interview sequence within that division, resulting in C1 through to C4. This was repeated for service providers SP1 to SP7, and in-house being IH1 through to IH8. These allocations were used as labels to delineate the different recordings as opposed to guiding a sequential analysis in that order.

The initial stage was to assess the research pool in terms of their biographical details to assess how this represented FM (as per Chapter 2), and this is illustrated in Table 5.3. A further breakdown of this is detailed in Appendix 4.

This would indicate a good sample spread across the industry. It was slightly male dominated (11:8 ratio), with the age range being thirty-six to fifty-five. The majority length of service was over fifteen years, indicating

Table 5.3 Participant details

<table>
<tr><th></th><th>1</th><th>2</th><th>3</th><th>4</th><th>5</th><th>6</th><th>7</th><th>8</th><th>9</th><th>10</th><th>11</th><th>12</th><th>13</th><th>14</th><th>15</th><th>16</th><th>17</th><th>18</th><th>19</th></tr>
<tr><td>Gender</td><td colspan="11">Male</td><td colspan="8">Female</td></tr>
<tr><td>Age</td><td colspan="2">26–35</td><td colspan="8">36–45</td><td colspan="7">46–55</td><td colspan="2">56+</td></tr>
<tr><td>Tenure</td><td colspan="4"><5</td><td colspan="3">5–10</td><td colspan="3">11–15</td><td colspan="9">15+</td></tr>
<tr><td>Typology</td><td colspan="7">Service Provider</td><td colspan="8">Inhouse</td><td colspan="4">Consultant</td></tr>
<tr><td>Sector</td><td colspan="8">Public</td><td colspan="11">Private</td></tr>
<tr><td>Delivery</td><td colspan="4">Hard FM</td><td colspan="7">Soft FM</td><td colspan="8">Total FM</td></tr>
<tr><td>Level</td><td>Dir</td><td colspan="5">Executive</td><td colspan="9">Senior Mgmt.</td><td colspan="4">Mgmt.</td></tr>
</table>

Source: authors.

that it is a sustainable career path. There were seven service providers to eight in-house providers, and four consultants. This is in line with the analysis of FM in Chapter 1, which presented the discipline as 'one of the most outsourced components in organizations' (Ikediashi et al. 2014: 472).

The private sector has a slight lead over public sector representation (8:11). There was no presence from FM within the third sector. Total FM had a slight lead over soft FM, and hard FM had the least declarations, with only four participants. The main level was at senior management level (nine), but there was representation at all levels.

In line with the deconstruction phase approach, all nineteen research interviews were separated into their FM sub-categories of consultant, service provider, and in-house provider. These were then mapped across in accordance with the fairytale, ATU sub-category, and story typology as charted in Table 5.4. This allowed the dominant fairytale that represents the storied space in each division to be identified, providing the focus for the analysis.

5.4 Coda

This chapter has explained the development of the empirical stages of research. We have noted how the focus for analysis centres on the three main FM service delivery mechanisms defined—the professional consultant, the external service provider, and the in-house function. We have described how, initially, the tale of *Rumpelstiltskin* forms the basis for a pilot deconstruction analysis—one facilitating casting across the four main characters of the story. Subsequently, the main issues raised from a collaborative research approach to storytelling are described before, finally,

Table 5.4 FM divisions mapping to fairytale typologies

No.	Fairytale	ATU no. and sub-category	Story typology
Consultant:			
C1	*Little Red Riding Hood*	333 supernatural opponents	Overcoming the monster
C3 C4	*Hansel and Gretel*	327 supernatural opponents	
C5	*Snow White and Seven Dwarfs*	709 other stories of supernatural	Rebirth
Service Provider:			
SP1 SP3 SP6	*Goldilocks and Three Bears*	531 supernatural helpers	Voyage and return
SP2	*Rapunzel*	310 supernatural opponents	Overcoming the monster
SP7	*Little Red Riding Hood*	333 supernatural opponents	
SP4 SP5	*Cinderella*	510 supernatural helpers	Rags to riches
In House:			
IH1 IH4 IH6 IH7 IH8	*Cinderella*	510 supernatural helpers	Rags to riches
IH1 IH5	*Puss in Boots*	545 supernatural/enchanted relatives	Rebirth
IH3	*Beauty and the Beast*		

Source: authors.

an interview-based analysis—gathering individual lived-experience stories and forms of creative scenario re-imagining—is outlined. The latter approach identifies narratives to be assessed in line with the three dominant fairytales emerging in this phase of the work—the consultants' *Hansel and Gretel*, the service providers' *Goldilocks*, and the in-house function's *Cinderella*.

From the original ten fairy stories (Appendix 2), only two were not selected during the interviews, *Jack and the Beanstalk* and *Sleeping Beauty*, although their story arcs are represented in the *supernatural opponents/overcoming the monster* and *supernatural relatives/rebirth* categories. To support the analysis of the focal tale in each division, the three fairytales are now analysed.

6
The Trinity of Tales

As described, in the research process the analysis each of three service streams (consultant, service provider, and in-house) of FM was related to a different dominant fairytale. To effectively interpret the tales, a fuller analysis of each chosen tale is now undertaken. In this chapter, the three tales of *Hansel and Gretel*, *Goldilocks*, and *Cinderella* are retold, using the Ladybird classical series as a guide. This then allows a detailed analysis of each tale, encompassing the interpretations and evolutions, and providing a final review of where the tale has surfaced within the current FM literature.

6.1 *Hansel and Gretel*

Hansel and Gretel are siblings that lived on the edge of a forest, with their poor woodcutter father and stepmother. With food running out, the woodcutter laments their dire situation, but the stepmother has a solution: 'I know what we must do' she said. 'Tomorrow we will take the children into the forest. Then we must leave them there. They will never find their way home. We shall be free of them!' (Southgate 1964: 6). The woodcutter is initially appalled, but she persists, insisting that he is a fool to let all four of them die of hunger. 'The woodcutter's wife gave him no peace, until at last he agreed' (1964: 6).

Too hungry to sleep, the children overhear this and Hansel sneaks out of the house and fills his pockets with white pebbles. In the morning, the stepmother takes them to the woods to gather firewood, and so Hansel drops the pebbles as they go. The father lights a fire for them, telling them to stay as he and the stepmother go to gather more wood. They fall asleep, and when they wake up, they are alone. By following the pebbles that shone in the moonlight, they return home, much to their father's delight as 'it has broken his heart to leave them in the forest' (Southgate 1964: 14). The stepmother was not so pleased and scolded them for getting lost.

Once Upon a Time in Facilities Management. Patricia McCarroll and John Hassard, Oxford University Press.
 DOI: 10.1093/9780191943492.003.0006

Before long, food was scarce again, and the stepmother once more suggested getting rid of the children and the woodcutter again submitted to the plan: 'he would rather have shared his last crust with his children. But his wife would not listen to his pleading, and again, he had to agree' (Southgate 1964: 16). However, this time she had locked the door and Hansel could not get out to collect any pebbles. Instead, he left a trail of breadcrumbs as they were led into the woods. When they awoke, this time to find themselves alone, the birds had eaten the breadcrumb trail and they got lost deeper and deeper in the forest.

Just as the children were at risk of dying from hunger and fatigue, a little cottage appeared that was 'made of bread and cakes, and the windows . . . made of sugar' (Southgate 1964: 24). As they were eating at the house, an old friendly woman appeared on crutches and invited them in for a safe night's sleep. The next section of the story deserves an excerpt from the Ladybird edition, not only revealing the major plot line but also the succinct literary nature of fairytales:

> Hansel and Gretel did not know that the old woman was really a wicked witch, who trapped children. She couldn't see very well, but she had a fine sense of smell. She could smell children coming. The house of bread and cakes had been built to tempt children in. The witch gave an evil laugh. 'These two shall not escape,' she cackled. Early next morning, the witch pulled Hansel from his bed, and locked him in a cage. Although he screamed, there was no one to help him. (Southgate 1964: 28)

Gretel was made to clean the house in a slave-like manner and prepare huge dishes for Hansel, as the witch was trying to fatten him up in order to eat him. Every morning, she ordered him to hold out a finger so she could feel if he was fat enough to eat. Hansel, knowing the witch had poor eyesight always held out a bone. Eventually, the witch became tired of waiting for him to gain sufficient weight and decided to eat him anyway.

Planning to also eat Gretel, the witch ordered her to get into the oven and check it was hot enough. Gretel, having guessed the witch's plan, creates the excuse that she is too big to fit into the oven. '"You silly child", the witch said. "Look, I could even get in myself!" She bent down and put her head into the oven. Gretel gave her a hard push, and she fell right inside . . . Gretel ran to Hansel's cage. "The witch is dead!" she cried. "We're safe. Now I must get you out of that cage."' (Southgate 1964: 36).

Searching the gingerbread house before leaving, they filled their pockets with pearls and precious stones and went out into the forest again. Although it was dark, friendly birds and animals guided them until they found their way home. Here, they were reunited with their father who was overjoyed to see them, as he was now alone as the stepmother had died. Using the money from the jewels, they all lived happily ever after.

6.2 The Breadcrumbs

Hansel and Gretel, like *Goldilocks*, and *Cinderella*, is set in a family dynamic which represents a stable situation that is thrown into a state of disequilibrium by the addition of a new member (Tatar 1987). The protagonists are abandoned by the very people put in place to protect them, and as such the tale 'raises problems that pertain to the consequences of hunger in poor families, the trauma of abandonment, the depiction of women as nasty stepmothers and witches, survival and the sanctification of paternal rule' (Zipes 2006: 195).

Historically, this tale may have originated from a great famine in the early fourteenth century that spread across Europe. In the United Kingdom alone, it is estimated that 10–15 per cent of the population died from hunger and disease (Sharp 2013), with crimes such as cannibalism and infanticide being evident. Reflecting times of extreme hardship, this tale has an ATU categorization of *supernatural opponent*, indicating an omnipotent malevolent presence that must be destroyed against all odds to secure a happy ever after. As such, the tale of *Hansel and Gretel* is a story of hunger and hope, cruelty and courage with a final victory creating a 'soothing, pacifying tale that touches on issues of abuse and abandonment and provides hope that security and happiness can be found after a traumatic episode' (Zipes 1997: 58).

Hansel and Gretel appears in all the Grimm Brothers' editions, with little alteration in the overall story arc. The main difference lies in the mother role, who was declared as the biological parent in the 1812 first edition, but by the seventh edition in 1857, this character became a stepmother. This alteration to a stepmother remains to the present day and is an adaptation to soften the tale, as the thought that a mother would abandon her own children was considered too upsetting for a fairytale, as they became increasingly associated with children (Zipes 2014a). However, due to the high rate of women dying in childbirth in the medieval age, the presence of stepmother and stepsiblings were common in households (Zipes 1979). Perrault, C. 1697/2020 does not record this tale, although the story arc supports his version of Tom Thumb, where seven sons are abandoned in the woods by their parental figures (Tatar 2002).

The other main difference between the Grimm Brothers' version and that of the more contemporary tale is the inclusion, in the latter, of the jewels that the protagonists find in the witch's house. However, the Grimm Brothers' first edition text does conclude the tale stating that the woodcutter was a rich man. It does not give any further information but concurs that the mother had died.

Overall, this short tale, containing five characters, has clear links to an initiation story and therefore a connection to liminality. The protagonists are separated from their familial home and abandoned into the liminal state of the woods, going on a journey in order to be accepted back into the household. Knowledge of survival skills, twinned with the riches they acquire, serves to change their roles, and wins them acceptance back into the family home. They leave as vulnerable children and return as victors over evil—crucially with resources to ensure the group's ultimate survival.

Turning attention to the individual characters, the sibling protagonists are indicative of the hero archetype. Hansel saves them both with his plan of laying trails of breadcrumbs and pebbles, with Gretel then saving them both by her actions within the gingerbread house. In this tale there are no gender roles associated with the genre of fairytales, for both male and female characters are equally clever and survive by outwitting those who would threaten them, that is, all the other people in the story. They have no supernatural help, but rather a *supernatural opponent,* as indicated by the classification of the tale within the ATU scheme (tale number 327) and alignment to *overcoming the monster* narratives (Booker 2004). The symbolic monster is surviving their greatest fears, represented by children who are abandoned by their parents and exposed to predators (Tatar 2002).

The stepmother character is only present at the start of the tale. Her death, which occurs when Hansel and Gretel battle the witch, is not elaborated on but is key to the happy return of the protagonists. This role is generally portrayed as a malevolent character, as it is at her insistence that the father abandons the children in the woods, with 'poverty providing the basis for the hardness of heart' (Lüthi 1976: 64). However, the children still return home the final time, with the intention of sharing the jewels with everyone, being unaware that the stepmother has died in their absence. This would indicate that the protagonists have no ill will towards this character, despite her actions against them, allowing the possibility that they 'reluctantly comprehend the situation which forces their parents to act as they do' (Zipes 1979:

38). The happy ending of the tale is therefore formed in the creation of a contented and stable family.

The father is also a side character, appearing in the tale only at the beginning and the end. He is described as very poor and unable to feed his family. Ultimately, the father is harassed into the abandonment of the children, bending to the force of his new wife. This is true for most of the retellings of *Hansel and Gretel*, where 'the mother stands alone as the villain' (Tatar 1987: 36). In some readings, therefore, the father is also the victim: long suffering and stressed, and strangely committed to at least fulfil his agreement to his partner, for 'he who has begun a thing must go on with it' (Grimm Brothers 1857, as quoted in McGlathery 1991: 5).

Symbolically the father arguably represents the safe home (Zipes 2006), and as such he is not held responsible for the abandonment of the children. In fairytale fashion, the children happily journey home twice. In the final return, the riches the children bring from the witch's home means the father will ultimately benefit from the children's actions, this carrying with it the symbolism that the 'acquisition of riches or money will solve all their problems' (Zipes 2006: 205).

The villain in the tale, the supernatural opponent who must be overcome to ensure the survival of the protagonists, takes the form of the witch who lives in the gingerbread house within the woods. In comparison to the active role of supernatural opponents such as the wolf in *Little Red Riding Hood*, the witch is passive, awaiting the victims to visit her lair by displaying a deceptively harmless and charming character of a little old lady. However, this villain is heartless, representing a role whose only concern is to 'look after its own interests at the expense of everyone else in the world' (Booker 2004: 32). The wickedness of this role is emphasized by her cannibalism, a taboo that is heightened by the representation of children as lost and innocent. The supernatural opponent, as manifested in the witch, has therefore been aligned to 'the entire feudal system or the greed and brutality of the aristocracy, responsible for the difficult conditions' (Zipes 1979: 38), with this relating to the widespread famine at the time of the story's origins.

The strong characters and story arc are supported by striking imagery throughout the tale. Being lost in the dark woods indicates confusion and separation, symbolic of a liminal state. The phantasmagorias of the pebbles and breadcrumbs are both white, the colour of innocence and naivety. Set in times of famine, the breadcrumbs are also a precious resource and common symbol, indicating the idea of being lost and therefore representing a fragile and insecure state. The breadcrumb trail is only used in times of great desperation, with no guarantee of leading the main characters back to safety, or else them on a different journey to the new and unknown. The other memorable imagery from the tale is that of the gingerbread house, which appears magically in the forest and at the point of most need for Hansel and Gretel. However, the sweet exterior is a veneer, and the gingerbread house turns from place of salvation and joy into one of confinement and desperation.

Hansel and Gretel remains one of the most popular fairytales in contemporary culture, as an allegory of how it is possible to 'survive in an unjust society and struggle with hope' (Zipes 1979: 39). It has been republished many times within Ladybird editions for children, being the subject of many operas and re-imagined by contemporary authors writing for both youth and adult audiences, such as Naomi Mitchison in 1936, Roahl Dahl in 1989, Michael Faber in 2006, Neil Gaiman in 2014, and Philip Pullman in 2015. 'The wonderful invention of the edible house, together with the implacable cruelty of the witch and the wit and bravery of Gretel in dealing with her so neatly, make [the tale] unforgettable' (Pullman 2012: 85).

In FM, the concept of *Hansel and Gretel* has appeared in reference to the management of health-care facilities and the merging of different clinicians and specialisms within it (Carpman and Grant 2016). This pulled specifically on the imagery of breadcrumbs, to find a clear path in large change programmes. Exadaktyloas and Velmahos also deployed the tale as an analogy in similar health-care circumstances. In this research it was observed that 'teamwork, optimism, and some good luck, emerges in this classic German folktale; the same timeless values are imperative for those involved in the ongoing struggle for change, improvement and the creation of new concepts in patient care' (2008: 321).

6.3 *Goldilocks and the Three Bears*

The tale of *Goldilocks and the Three Bears* starts with a description of the latter, with the father bear being very big, the mother being medium sized and the baby being very small. The mother bear prepares breakfast porridge for the three bears in accordingly sized bowls, at which point they go for a walk to let it cool. Goldilocks, who has long, golden hair, was walking in the woods and came across the three bears' little house. She could see the door was open and that no-one was at home, so let herself in. Having not eaten breakfast that morning, she was hungry and so tasted all the porridge. The first one in the large bowl was too hot, the second in the medium-sized bowl was too lumpy, but the one in the smallest bowl was just right and so she ate it all. Moving round the house, she came across three chairs. The first one was too high for her to sit in, the medium-sized one was too hard, but the smallest one was just right. However, she was too heavy, and it broke under her weight. Getting sleepy, Goldilocks then went into the bedroom, where the big bed was too hard, the medium too soft, but the smallest just right, and so she climbed in and feel asleep.

The bears then returned from their walk to eat their breakfast. In turn, they all ask, 'who has been eating my porridge?' (Southgate 1969: 34), with Baby Bear exclaiming that his had been eaten all up. They discover the chairs and question who had been sitting in them, with Baby Bear 'in a tiny, little voice' saying 'who has been sitting in my chair and broken it?' (Southgate 1969: 34). Moving to the bedroom, they ask in turn, 'who has been lying on my bed?' (Southgate 1969: 34), until they reach Baby Bear's bed: 'Here she is, cried baby bear, making his tiny, little voice as loud as he could. Here is the naughty girl who has eaten my porridge and broken *my* chair! Here she is!' (Southgate 1969: 47, italics original). This woke Goldilocks, and she jumped out of the bed in fright, then jumping out of the window and back into the woods. She was already out of sight by the time the bears reached the window, and they never saw her again.

6.4 Just Right

Of the original tales as selected by the FM workshop (Appendix 2), this one is an oddity, as it does not appear in either the collections of Perrault (1697) or the Grimm Brothers (1812–1857). However, it does have similarity with *Snow White* (which is in all the Grimm Brothers' editions), when the lost princess enters the seven dwarfs' home, eating their food and falling asleep in their bed (Opie and Opie 1980). As the story of *Goldilocks* was not included in these seminal works, it has arguably undergone the most change before stabilizing in the current recognizable form as illustrated above.

The literary version of this tale is attributed to Robert Southey, in his 1837 collection of essays, and specifically an entry entitled 'Story of the Three Bears'. It also appears later, in 1849, in Joseph Cundall's collection of tales for young children (see Opie and Opie 1980). However, a less well-recognized version of the tale is found in 1831 in a book Eleanor Mure created for her nephew's fourth birthday, in which she scribed all the tales she was familiar with from the oral tradition (Tolovaj Publishing House 2020). In these earlier versions, the main protagonist is not a young girl with the striking golden hair—for Cundall's 1849 version, for example, has her as a young girl with silver hair, a possible connection to the grey hair of the old woman in the earlier tales. It is not until the turn of the century where the hair becomes golden—with the full title of *Goldilocks* appearing in Hassall's 1904 'Old Nursery and Rhymes' collection. Reflecting the lead protagonist altering in appearance, both the role of the main character and the story arc have mutated to form the tale's current ambiguous state. For instance, when the tale had an old woman as the protagonist, there was a clearer link to this being a villainous character. In Mure's 1831 version, however, she is a witch-like character, who does not escape into the woods at the end but is impaled on a church steeple. Where the protagonist is portrayed as a young girl, she remains villainous in the fact she represents a naughty and rude child, an intruder who breaks into private property, damaging and stealing. The consequences of this are demonstrated when she must face the three bears, but ultimately, she escapes free of bodily harm into the woods. The portrayal of the bears as a family unit empathizes their wholesomeness, allowing the dichotomy of good/bad to operate in relation to the four characters.

This tale is indicative of the use of the 'rule of three' in fairytales (Booker 2004). With the family unit consisting of three bears, it follows that there are three chairs, three bowls of porridge and three beds that are tried in succession by the protagonist. The rule of three has also led to many interpretations of the tale's meaning, with the repeated actions of Goldilocks highlighting her role as an intruder who cannot control herself, especially when dealing with items that do not belong to her (see Tatar 2002).

If viewed as a cautionary tale, it highlights the risks of getting lost and wandering away into the unknown, with the repetitive formula used to reinforce the increasing danger to the protagonist, who is unaware of the peril presented in this strange environment. In this view, Goldilocks is not a villainous character but one more representing naivety, as she is seemingly unaware of the damaging nature of her actions.

Therefore, the protagonist role of Goldilocks is either one of innocent naivety or damaging intruder. What remains is the role of the three bears, who reflect a stable family unit whose house and possessions have been violated by Goldilocks, emphasized by the impact on Baby Bear, the protagonist's peer and whose possessions are therefore affected the most. As the structure of fairytales dictates surface details, there is no overt aggression described in the tale, although the presence of the bears scares Goldilocks sufficiently that she jumps out the window to escape.

In this way, the role of the three bears is that of the *supernatural helper*, as indicated by this tale's ATU classification (tale number 531). The bears are supernatural due to their ability to talk, and ultimately are cast into the role of helper, allowing Goldilocks to learn respect for other people and their property. The ambiguity of the tale is highlighted using Booker's (2004) category of *voyage and return*, which casts the focus back to Goldilocks as the protagonist, and her specific journey as told within the tale. Once again, as with the tale of *Hansel and Gretel*, themes of metamorphosis and liminality are apparent. Goldilocks ventures from home into the dark woods (separation) and finds the house of the three bears. The initial excitement she feels while exploring is replaced by growing tension as the bears return (liminal state), ending with her escape (aggregation). Goldilocks is allowed to return, presumably back to the safety of home, but as a wiser and more cautious character (Booker 2004).

Later interpretations focus on the role of satisfaction contained within the three actions, linking to the *just right* ideology seen in creation of term the Goldilocks Principle or Goldilocks Effect (Straker et al. 2018). Found in fields as diverse as psychology, biology, economics, engineering, and astronomy, the term represents the element of perfection. The imagery has been specifically used in FM to highlight the physical boundaries that the discipline often enforces within office settings. For instance, analysing FM's role in policing open-plan offices, and notably making unplanned audits of shared areas, can produce the reaction 'who's been sitting in my chair?' (Marmot and Eley 2000: 35).

The tale itself remains popular in UK culture, with the 'Story of Pretty Goldilocks' appearing in the Andrew Lang collections (1889–1913) and the story having many imprints within the Ladybird Classics collection (1867–present). Modern forms of re-imagining the tale have been found for example in popular works by Roald Dahl (1982) and Neil Gaiman (2014).

6.5 *Cinderella*

The story of *Cinderella* starts with a focus on the protagonist, who lives with her father and two elder sisters; their mother is dead. Although the siblings were all originally beautiful, Cinderella's sisters were bad tempered and unkind, so their faces grew ugly. Jealous of Cinderella's kind nature, they made her toil hard from dawn to dusk—fetching coal, cooking, and cleaning. Cinderella also had to dress her sisters in their fine clothes and brush their hair, while she remained in rags. 'In the evening, when she had worked until she was weary, Cinderella had no bed to go to. She has to sleep by the hearth in the cinders. That was why her sisters called her Cinderella and that was why she always looked dusty and dirty' (Southgate 1964: 6). Although sisters, everyone thought Cinderella was the maid.

The king of the region organized a feast for three days so his son could pick a bride. All the beautiful young girls in the country were invited and Cinderella's sisters could talk of nothing else in their excitement. Cinderella dressed her sisters on the eve of the first ball, while the latter mocked her for wanting to go to the ball in her rags. They left her crying by the fire, when Cinderella's fairy godmother appeared and asked her what the matter was. On hearing how much she would love to own a gown, go to the ball, and see the prince, the godmother replied, 'And so you shall, my dear . . . dry your eyes and then do exactly as I tell you' (Southgate 1964: 12).

From the garden, Cinderella had to pick the biggest pumpkin, which the fairy godmother turned into a carriage—one made of gold and lined with red velvet. The mousetrap was gathered from the pantry, and the six mice within it turned into fine grey horses to pull the carriage. The rattrap was brought from the cellar and the rat within it turned into a smartly dressed coachman. Lastly, two lizards were brought from the garden and turned into smart footmen. With her wand, the fairy godmother then touched Cinderella, who was immediately dressed in the finest pink ball gown and pink satin slippers.

Cinderella then set off for the ball, with the strict instruction that she must return home before midnight, or else everything would return to its original state. At the ball, no one recognized Cinderella in all her finery, and it was rumoured she was a princess from another country. The prince was overcome by her beauty and danced with her all night, not letting anyone else intervene. Despite having the best night of her life, Cinderella remembered the warnings of the fairy godmother, and slipped away at quarter to midnight. She returned home where everything reverted to normal, and she was once again dressed in rags. Sitting among the ashes, she listened in silence to the tales of her sisters, and specifically of accounts of a beautiful but mysterious princess who had danced with the prince all night.

On the second day of the ball, the godmother appeared again and transformed everything as before, but now Cinderella's dress was even more beautiful, being blue with matching slippers embroidered with silver. The prince again danced with her all night, and Cinderella almost forgot her curfew, hurrying to her carriage at five minutes to midnight. Halfway home, her attire and entourage transformed, leaving her dressed in rags in the middle of a dark and lonely road. Running home, she managed to sit herself by the ashes of the hearth again before her sisters came home.

On the third and final day, Cinderella's dress was the most magnificent yet, made of gold and silver lace, with golden slippers and a diamond tiara. Dancing once more with the prince all night, it was only when she heard the clock start to chime midnight that she remembered her fairy godmothers' warning. Terrified she would transform into rags in front of her prince, she ran as quickly as she could from the crowd, losing her left slipper in her haste. With the carriage gone, she ran all the way home.

Finding the slipper, the prince went to his father, the king, saying, 'no one shall be my wife but she whose foot this golden slipper fits' (Southgate 1964: 42). And so the king's herald, led by the prince, toured the streets of the city as every lady at the feast attempted to squeeze a foot into the slipper, but it was too dainty for anyone who tried. The last house was Cinderella's, where her sisters tried so hard to make the slipper fit, they made their feet bleed.

> At last, the prince turned to Cinderella's father and asked, 'Have you no other daughter?'
>
> 'I have one more', replied the father, 'but she is always in the kitchen.' At this point, the ugly sisters cried out, 'She is much too dirty, she cannot show herself.' But the prince insisted and so Cinderella was sent for. (Southgate 1964: 46)

As the slipper fitted Cinderella, the prince looked up into her face and recognized his princess. At this moment, the fairy godmother appeared and transformed Cinderella's rags into a velvet gown, with the prince subsequently placing Cinderella on his horse in order to return to the palace to arrange their wedding. The sisters were horrified to find out the beautiful princess at the ball was their dirty sister—indeed 'so angry that they were pale with rage' (Southgate 1964: 50). Cinderella and her prince then lived happily ever after.

6.6 The Glass Slipper

This tale is the most complex of the three selected stories, with a more developed plot, and as many as eight character types being declared. As with the tale of *Goldilocks*, it has an ATU category of *supernatural helper* (tale 510a), but here there is no ambiguity over association with the fairy godmother role. The Booker (2004) classification of *rags to riches* emphasizes the familiar story arc, which is the main metaphor associated with this tale, as literally portrayed in the description of the protagonist being transformed from a poor underling covered in tatters to an elaborately dressed princess.

Cinderella is considered one of the most popular fairytales in the world (Zipes 2013), with 'over 700 documented versions dating as far back as 9th century China' (Robinson and Wildersmith 2015: 55). As such, this tale appears in both the seminal works of Perrault (1697) and the Grimm Brothers (1812–1857). It was arguably Perrault who gave Cinderella her infamous glass slipper, possibly mistaking the word *vair* (fur) for *verre* (glass) in oral translations (Pullman 2012: 126). Historically, changes in details of the tale across versions and publications tend to reflect changes in the environment in which they were recorded. In Perrault's 1697 tale, the sisters are identified as being stepsisters, and they beg for forgiveness and so were welcomed back into the family (Carter 1977). They achieve their own happy-ever-after status by marrying other lords in the court. *Cinderella* was one of the eight tales transcribed for his renowned publication *The Tales and Stories of the Past with Morals* (*Histories ou contes du temps passé*—subtitled *Tales of Mother Goose*), and this version ends with a firm statement of the ethics of the tale. Indeed, here two interpretations are declared. The first concerns the physical beauty of Cinderella, which is represented in her outward appearance, while the second is a symbol of her personal qualities, notably in terms of charm and grace. It was these qualities that make her worthy to be a queen, with the second moral being elaborated as follows:

> *You have a great advantage, I admit,*
> *If you received from Heaven at your birth*
> *Good breeding, courage, sense, a ready wit,*
> *And other things of comparable worth;*
> *But that is not enough, unless you know*
> *How best to use such precious gifts: you need*
> *A godfather or godmother to show*
> *What you must do in order to succeed.*
>
> **(Betts 2009: 141)**

As stated in Chapter 3, all fairytales are cultural memes reflecting the wider environment at the time they were committed to paper, and it is worth noting here Perrault's position as a political aid in the court of Louis XIV. Therefore, Cinderella in this tale, is arguably positioned as an exemplar of 'proper, civilised behaviour for the French noblewoman to emulate' (Zipes 2006 as quoted in Robinson and Wildersmith 2015: 55).

Perrault (1697) identified the roles of stepsisters, but the earlier versions of the Grimm Brothers' *Cinderella* (1812) identify them as full relations which is altered to stepsiblings in later editions. In the Grimm Brothers' first version of the Cinderella story (1812), the sisters suffer in the tale as they cut sections of their feet trying to make the slipper fit. However, the stepsisters were overtly punished from the 1819 version onwards, where they went to the wedding to share in Cinderella's happiness, but on the way to the church doves peck out their eyes (Tatar 1987). This cements the sisters as Cinderella's opponents because they will do anything to ruin Cinderella's chances for happiness in order to promote themselves. The idea that these are the 'ugly' stepsisters' echoes Perrault's morals of beauty being a quality rather than an aesthetic commodity (Booker 2004).

The father character plays a similar role to that portrayed in the earlier tale of *Hansel and Gretel*. He does nothing to protect his offspring, thereby 'contributing to . . . suffering through benign neglect and abandonment' (Zipes 2006: 115). Where Hansel and Gretel's father illustrates remorse, this is not alluded to in *Cinderella*, and the father character disappears from the main text as the story proceeds. He does not benefit from his child's new wealth, but neither is he punished like the stepsisters. The other father figure in the tale, the king, is also sparsely mentioned in true fairy story structure, other than ordering the balls to take place for his son to find a wife, and then blessing the union.

Cinderella's character is that of an abused and unwanted child, but one who represents a survivor. She 'is helpless, forsaken and alone, like so many other heroes and heroines in the fairy tale' (Lüthi 1976: 61). The start of the tale illustrates that Cinderella accepts her role with humility and endurance and bears no overt ill will to the family that mistreats her. Here, the character is 'prototyped as a sweet, modest, attractive girl, who works diligently and endures hardship from the hands of others with both patience and grace' (Robinson and Wildersmith 2015: 60). This is evidence of the ability of the fairytale be symbolic (Holbek 1987), where the real world is referenced through fantastic events, objects, and emotional impressions. The beautiful ball gowns are markers of Cinderella's real nature and values, and the prince selects her by subconsciously understanding these signs (Toomeos-Orglaan 2013). The character of the prince symbolizes the riches in opposition to Cinderella's rags, but along with financial wealth the prince also represents recognition of and love for the protagonist.

In terms of other themes, the three nights that Cinderella must journey through are once again indicative of the 'rule of three' within fairy stories, as also illustrated in *Goldilocks and the Three Bears*. Additionally, although aided by her supernatural helper, the magic has its limitations and Cinderella must resist temptation and leave by midnight. This alludes to the protagonist's ability to uphold promises, obey rules, and avoid short-term gains in favour of long-term benefits. Dégh analyses the transformation of Cinderella as part of a learning process, one giving time for her to 'become stronger, cleverer and more beautiful' in symbolism that 'also serves as a sign of the protagonist's "maturation"' (1994: 94).

The role of Cinderella has also provoked debates on the passive nature of the character, as she reacts to situations around her as opposed to leading them. This idea has spawned the theory of the Cinderella Complex, a phrase coined by the book of the same name published in 1981 by Colette Dowling. This refers to 'a network of largely repressed attitudes and fears that keeps women in a kind of half-light, retreating from the full use of their minds and creativity. Like Cinderella, women today are still waiting for something external to transform their lives' (Zipes 2006: 112). However, a counterargument is that Cinderella is empowered by being able to endure her abusive early conditions and does not hesitate when given opportunity.

Once again, a strong theme of liminality is apparent, although overtly made through the change in the protagonist's status. The start of the tale portrays Cinderella as an outsider ostracized by the family unit: 'not many people had ever been nice to her, and she was a stranger to the feeling of being popular and sought after' (Pullman 2012: 120). Thus, the royal balls she attends, and new social and emotional journeys she undertakes, serve to shape a new identity, until she finally emerges as the queen, and one fully incorporated into kingdom life.

However, this transition is not something she could have achieved without supernatural help, which manifests itself in the role of the fairy godmother. This role does not appear in the first version of the Grimm Brothers' retelling, in 1812, where the supernatural helper is the spirit of Cinderella's deceased mother. In fact, there are no overt morals stated in any of the Grimm Brothers' editions, although the increasing magical element of the fairy godmother is controverted by the escalated punishment inflicted on the sisters, because of their treatment of Cinderella. Additionally, the transformation of rats and lizards to coachmen and footman at the wave of the fairy godmother's wand, alongside the image of the infamous pumpkin carriage, is a direct link to Mother Nature, for 'Cinderella may be subjected to disgrace and humiliation by a stepmother and her daughters, but nature comes to her rescue' (Tatar 1987: 73).

The story of Cinderella therefore is ultimately a story of transformation, the 'representation of that dirty humiliated girl who proves herself to be beautiful and a winner/survivor despite all the ashes and cinders that are heaped upon her. We recognize her for what she is—a true princess'

(Zipes 2006: 107). The tale contains strong imagery—which has ensured the endurance of the tale, and its multiple reinterpretations and uses, in modern society. For instance, the increasingly violent punishment of the stepsisters seen in the Grimm Brothers' editions 'may be a useful antidote to the sentimental modern view, which encourages you to live with almost anything you can find. What the envious girls found was not good; the punishment fits the crime' (Lüthi 1976: 14). Its *rags to riches* categorization (Booker 2004) is one of the most recognizable archetypes in contemporary society, with 'young men and woman emerging from humble anonymity to receive obsessive adulation as film stars, pop singers, supermodels, sporting heroes' (Booker 2004: 585), symbolism revealing their natural glory and balanced self.

The extended metaphor of Cinderella has also been used to explore organizational issues, as demonstrated in Harris et al. 2013, which used the tale to explore gender advancement in academia. Within the specific field of FM, this story has an overt connection as it is associated with being a 'Cinderella service', for 'FM goes on "behind closed doors", its practitioners doing all the preparatory stuff that ensures everyone else can enjoy themselves at the legally compliant, sustainably resourced and perfectly organized Ball' (Read 2016, online), with this being largely a reference to catering and front-of-house activities.

The Cinderella symbolism also extends to the physical placement of FM services in many organizations, with this connecting to the imagery of Cinderella sleeping by the hearth. Becker refers to this as 'the corporate basement—where facility management was often housed symbolically, if not literally' (1990: 49). Although, the symbolism of FM with the *Cinderella* tale is not complete, for it aligns to the first stage of the story before the intervention of the fairy godmother—thereby reflecting negative connotations—the use of this imagery is beginning to alter in some studies. Marmot and Eley report, for example, that 'in large UK companies surveyed recently, none had its most senior property [FM] executive on the main board, although they generally agreed that these days the property department is a bit less of a Cinderella' (2000: 61).

6.7 Coda

As seen in Chapter 5, the three chosen fairytales fall within different analytical categories. Each is retold in line with the most popular contemporary version—notably as represented by alignment to the Ladybird Classics series.

This allowed a detailed analysis of each fairytale to be undertaken, considering the evolution of a tale from its oral to literary versions. In so doing, uses of the tales in the FM literature were researched and assessed. By so detailing these tales, differences and similarities between them are shown—with this analysis revealing short yet dramatic tales that hold recognized symbols: symbols analysed in the next chapter in terms of the 'storied spaces' of the context and functions of FM.

7
The Storied Spaces of FM

As analysed in Chapter 6, the three principal FM tales detailed—*Hansel and Gretel*, *Goldilocks*, and *Cinderella*—hold different story categorizations. *Hansel and Gretel*, as chosen by the consultants, represented a storied space of having to *overcome a monster* (Booker 2004), with this being mapped across to the ATU category of *supernatural opponents*. In contrast, both the service providers and in-house providers chose ATU categories with a *supernatural helper* story arc, that is, *Goldilocks*, and *Cinderella*, respectively. However, they differ when Booker's (2004) categorizations are applied, revealing for example a *voyage and return* emphasis in the ambiguous tale of *Goldilocks* and an ultimately more positive association in the *rags to riches* motif of *Cinderella*.

This chapter focuses on conducting analysis within these overarching typologies. A comparative approach is applied, using the chosen fairytale from each service stream as the consistent event from which to describe, interpret, and analyse the different perspectives of the individual FM practitioners (Saunders et al. 2016). Previous themes such as the fluidity of roles, metamorphosis, liminality, relationship management, and enchantment, were used as guides throughout the comparative analysis. This was supported by the lived-experiences stories, as there was substantial overlap in many cases. For clarification, direct quotes from each interview are included, with these being referenced by a shorthand label (e.g. CI = consultant 1) declared in italics.

This method allowed established patterns of FM storied spaces, and their associated archetypes, to be identified across the three divisions representing the practice of FM, culminating in a ghostwritten fairytale being developed for each. With Chapter 2 identifying FM as an emerging practice, these ghostwritten fairytales capture the collection of idiosyncratic characterizations of *happenings* that constitute this organizational function, focusing on the plot and character castings of a BME story to form fairytales from FM, about FM. These detailed stages of critique and interpretation were repeated for each of the three divisions of FM, ensuring a consistent approach.

Once Upon a Time in Facilities Management. Patricia McCarroll and John Hassard, Oxford University Press.
 DOI: 10.1093/9780191943492.003.0007

It is noted in this approach that the identification of motifs and cataloguing is no longer so current in folklore studies, often being viewed as a 'positivistic fiction', and one reflecting the argument that 'we could know the meaning or history of a story by charting its variations in context and form' (Lindahl 1997: 263). Instead, they are used within this study in their current, temporal form in the recognition that they can instead 'promote accuracy of terminology and can act as keys to unlock large inaccessible stories' (George 1997: 205).

7.1 The FM Consultants' Storied Space: *Hansel and Gretel*

The tale of *Hansel and Gretel* emerged as the dominant story for the FM consultant category (illustrated in Table 7.1). First published in 1812, it remains one of the most popular fairytales in the world today (Zipes 2006). This fairy story falls within the ATU category of *supernatural opponents*, which is mapped to the *overcoming the monster* story plot, where there is a supernatural adversary that the protagonist must confront and defeat. The plot contains five characters: the protagonists Hansel and Gretel, their father and stepmother, and a witch.

As analysed in further detail in Chapter 6, *Hansel and Gretel* is a family-centric tale, where a stable unit is thrown into a state of liminality by the addition of an external entity (Tatar 1987), represented in this tale as the stepmother. Although the supernatural opponent is identified as a witch in this tale, there is also a villainous association with the stepmother. This character casts the children into the woods twice to starve, to save herself and her husband. As such, this tale has a strong theme of abandonment, with the protagonists being rejected by the very people who were there to protect them. However, this desertion places Hansel and Gretel on a journey that forces them to discover strategies for survival (Tatar 2015), allowing their return home and acceptance (Czarniawska and Mazza 2003).

It was this connection to being a family unit that led to this tale being chosen as the dominant one in the consultant division. C3's lived-experience tale concentrated on her role in attempting to rescue a struggling FM transformational-change programme in a local authority. In this tale, FM was seen as the family, bonded together under a shared title, but one which had become dysfunctional, with no leadership:

> So there's no control about who is doing what with whom, how it's being done, stuff is chaos. You find out things that have happened that you don't really know

> about, massive sums of money for example, for projects aren't really being monitored because, you know, Fred said to George, 'Oh yes, we can do that for you,' but there's no documentation behind it. (C3)

The family association was echoed by C4. This lived-experience tale relayed a space-planning project that C4 was brought in to correlate between the head office of the company and the different operational FM teams across a property portfolio. Here, the tale of *Hansel and Gretel* was instantly chosen, specifically concentrating on the role of the father as representing FM, and the stepmother being positioned with the main business:

> That's the one, [*Hansel and Gretel*] . . . that instantly hit me as . . . no matter what the facilities manager is going to say, the business is going to want this . . . and often the FM thinks, right, path of least resistance, what's going to cause less aggro . . . rather than putting the cards on the table, saying, yes, we could do that, however, let's do a bit more difficult work now and kind of challenge back what could be done for a better long-term future. So instead of leaving your kids in the woods, because you can't afford to feed them, let's put something in place now where we can build up our skill set, build up our sustainability, so we can include everybody. (C4)

Both C3 and C4's lived-experience stories, although detailing different scenarios, related to change programmes. One concerned an ongoing restructuring exercise—where C3 was appointed to rescue it when the main FM leadership team left post. C4's story related aspects of physical changes, notably the creation and merging of new offices, to the actions of staff working within these spaces. Here, he acted as broker between the largely operational FM function in charge of the buildings and instructions from head office. Both tales revealed much fluidity in terms of the range and timing of activities the consultants were commissioned to undertake. These stories revealed situations where many of the main decisions were already made and the programmes they were asked to deliver were already in difficulty. This left the consultants in a perilous position—working reactively and with little direction; effectively, lost in the woods.

FM consultants, therefore, are never truly integrated and therefore accepted in the commissioning organization. In FM, this analysis is a double-edged sword. Early works by Czarniawska and Mazza (2003) indicate that consultants by their very nature remain on the 'outside' of organizations, with this quality being accentuated by the liminal state of the FM profession as a whole. This imposed state of liminality would therefore indicate a masterfulness of metamorphosis. At a basic level, the consultant stories were both about change projects, and at a wider level the expectations of their roles would indicate that adapting quickly to change is an essential skillset within this FM division. In order to analyse aspects of relationship management, enchantment, and emerging archetypes, a more detailed scrutiny of the casting of *Hansel and Gretel* must be developed. To help guide this, a summary of the casting is provided in Table 7.1.

Traditionally, the tale starts by introducing the father, who has remarried. In many versions, he is referred to as a woodcutter, a symbolism picked up by C4:

> In the context of the book, this was probably a noble, valued profession. People need wood to burn. In FM, people need their buildings open, and warm, and cool, formal welfare facilities etc. . . . it is proving a valuable service to provide a certain standard, acceptable working conditions. (C4)

This is clear recognition of a legitimate role, and a sense of pride from the consultant in the role of FM in sustaining the organizational environment. However, despite having this valuable skillset, the family he is responsible for is suffering financially, in fact to the point of starvation. This lack of finance is echoed in both stories, reflecting the current 'age of austerity', with this being particularly highlighted in C3's tale of the public sector. Rather than embodying the archetype and deep role of a strong father figure leading his family, FM is portrayed as weak, and at the mercy of the loudest voice, which is seen most prominently in C3's identification to FM as a practice detailed above. Both stories associated this role with that of FM, placing the father as an embodiment of the FM industry or as a specific role given to the head of FM.

The father figure is perceived as being feeble in terms of external forces, but also in terms of self-awareness. By not realizing his own power, associated with the head-of-household role, this exacerbates the situation, forcing reactive and therefore disastrous situations to impact on others, which ultimately could be avoided by adequate forward planning. 'He is not recognized at all in terms of his importance, or what he is responsible for, [even though] he has brought these two children into the world' (C4), with no thought of how he was going to look after his responsibilities on any long-term basis. With this short-sightedness is an element of self-centeredness, possibly due to the vulnerability of a function and role that is so reliant on the main business for its survival: 'Stepmother is the main business. FM? If you keep it happy, it keeps you reasonably fed and clothed' (C4).

However, in both tales, FM is positioned as the father who causes harm by neglecting his duties; but does so unintentionally to a degree:

> FM is not the villain but is sometimes a little villainous to the users . . . but it's almost like a downtrodden character, [or] an understudy, wanting to be in the main cast. Probably got a good heart but is easily led to places they don't want to go. (C4)

Here the skillset of the practice is seen as capable, the practical woodcutting side, but the management side, that protects the resources necessary for survival, is still absent. Where for C3 this is associated with the negative deep role of the father (Moxnes and Moxnes 2016), for C4 a more material helper, in the form of a servant, is apparent. However, both cast the role of FM within a negative light, akin to the *shadow* archetype.

In the fairytale, the role of the stepmother can relate to that of the witch. Whilst not being supernatural, there is an argument that she mirrors the role of an opponent, for example, ensuring the children are cast out into the woods, in the understanding that they will most likely perish. However, in the re-imagining of FM, this character is more ambivalent. In C3's tale, this role is cast to senior management, who in this case was another consultant but without an FM background. In C4's tale, this role is affiliated to the main business, although the characteristics associated with this role suggest a picture of the stepmother as being an outsider. She represents an 'external force [that] has come into their lives and is telling him [the father/FM] what to do and [as such] she has to be listened to' (C4).

The stepmother sits outside the family unit, the perception of FM, and therefore does not truly understand it. Consequently, her decision to send the children to the woods is not in order to kill them, but is rather a pragmatic decision, one which reflects a lack of connection with the pre-existing group. This reflects the control of resources—together, they all die of starvation, but by sending the children out to fend for themselves, all four parties have an improved chance of survival. This is the *shadow* archetype:

> Is the evil aspect of the stepmother even real, or is this just the imagination of everybody else, as it's new, and it's different, because they are making painful decisions, but is that painful decision the best for the business as a whole? (C4)

There is also an undeniable element of power associated here with the stepmother. Not understanding the emotional dynamics of a family may justify her decision to cut the resources, but there was little evidence of dialogue with her partner. This would indicate a relationship formed under the understanding of monologue rather than dialogue, that is, of autocracy rather than true partnership. The role of the stepmother is not one to be challenged, reflecting a command-and-control (Price 2004) style that further emphasizes the powerlessness of FM as the secondary service.

The protagonists of the tale, those who the plot revolves around, are the siblings Hansel and Gretel. Portrayed as children, this conjures an image of them being naïve and defenceless, but their story is also one of survival in the face of abandonment. In the tale of *Hansel and Gretel*, the siblings are clever, and survive not though the intervention of a supernatural helper, as in Cinderella, but through their own bravery and ingenuity. Hansel provides the trail of pebbles (and later breadcrumbs) that forms a saving path, while it is Gretel who defeats the witch. As such, there is a theme of unity, of working together to ensure each other's survival in the testing situation they find themselves.

It was this bonding of the siblings, a continuation of the family theme, which appealed to C3 when developing a correlative analysis. Here, Hansel and Gretel were cast as the operational in-house FM staff, the ones who were most affected by the organizational restructuring that C3 was part leading. The positioning was not with the power of the children to overcome a desperate situation, but more that of the initial abandonment by those who were responsible for their well-being. For C3, the members of the FM department shared the family bond, where kinship to each other is clearly evident. They are the children who are 'trying to figure out what they are going to do, where their place is within this new kind of structure', for having been 'left very much to their own devices for their survival . . . they were literally making their own breadcrumbs' (C3). In this manner, symbolize the *adventurer* archetype, albeit reluctantly.

C4 also picked up on this symbolism and used the hierarchical nature of the family unit in explanatory analysis. With FM as an entity being the father figure, the characters of Hansel and Gretel were the end-users of a building, and as such relied on FM to provide them with shelter:

> FM has got their well-being or indeed productivity, but if the business (the step-mother) . . . tells FM to do something, FM is reluctant to do it, but will do it to appease the perceived level of power. End-users do have to look after themselves for survival. (C4)

As Hansel and Gretel were placed against their will into the woods, so C4 saw the end-users of office redesigns as feeling equally as out of place:

> Being put in the woods is like being put in an open-plan office. It is all noisy and distracting and it's 'no, can't do my job, this isn't helpful, I need to get out of here, what am I going to do?' (C4)

Like the siblings in the tale, there is a longing to return to what is familiar, to home, in times of distress. In terms of office redesign, for example, if the end-users are not happy, they will just:

> take it in their hands to make it work for themselves. If the office is open plan and everyone hates it in terms of noise and privacy, they will just put headphones in, going against the whole design concept. But it is them trying to get back . . . to go to what they know, how they do things. (C4)

Additionally, for C4, another side of Hansel and Gretel emerges: this manifests in negative childlike behaviour. Not recognizing the authority of parents who have directed them to leave, the siblings continue to return, placing the whole family unit at risk. Here there is an unwillingness of the end-users to accept the new situation and make it work for them—to form a new identity rather than reverting to the familiar, and yet unstable, old reality. This symbolizes the dark side of the *eternal child* archetype. The striking image of the breadcrumbs and pebbles are used to great effect here by C4, who acknowledges that FM handles the situation with no reference to the wider implications or enduring benefit:

> And those stones represent the old ways or something that links the user to the old way. Then the FM is going to pick up the stones and confiscate them and say, 'Sorry, you can't do that anymore, they are staying with me, you sod off back to the woods.' And they have to then go for something that was a little less robust that links back to the old way. So the breadcrumbs could be something that isn't often seen on the surface . . . like an attitude, that's not the way things are done around there . . . the breadcrumbs lead them back to the old world but they . . . get eaten up, and wind up in the gingerbread house of workplace ambiguity. (C4)

This again would indicate that FM discounts the management aspect of the profession, positioning to the *sage* archetype (Kostera 2012). Change-management elements were not considered, for a 'lift and shift' form of operational FM was applied in reality.

The final character is the supernatural opponent that forms the categorization of the tale, the witch. She is a passive character, lying in wait and then luring in her victims with unrealistic illusions. The witch is eternal and part of the natural order, 'the kind of unseen evilness that lurks in the woods that no one talks about . . . but catches you' (C3), aligning to the *shadow* archetype. Her portrayal as being a cannibal, of eating the projection of innocence that the children represent, highlights the evilness of this role, in opposition to the misdirection of the stepmother character. The recognizable symbolization of the gingerbread house was picked up in both re-imagining's, although represented in two very different interpretations. For C3, in the 'woods' of her change project, the gingerbread house represents that there are still good things out there, although sometimes they are hidden. For C4, a gingerbread house in the woods was a clear manifestation of how an environment can be manipulated, and he links this to the presence of ambiguity in the workplace. In terms of the witch who lives in the gingerbread house, she is cast not as a physical entity but as a manifestation of change, fear, and greed:

> I think the witch is possibly a big, dark cloud of fear, hatred, resistance to change, something cultural in there. It is more of a manifestation of something . . . their biggest fears, worries, maybe of temptation, a devilish character in the business . . . luring them in . . . Like, 'You don't want to work over there? Well, come and work over here, we will pay you a bit more . . . we will steal your soul and put a barcode on your forehead, but will give you more money.' (C4)

This also illustrates the vulnerability of FM consultants, and how reliant they are on the judgement of clients for survival. As a consultant, the uncertainty linked to a possible eternal liminal state may aggravate the evilness manifested by the witch. This reflects a state of uncertainty, of disconnection; of being cast aside for the next promise by an invisible entity which can make their client disappear overnight:

> It could also be, if the witch had got her way, and killed them and eaten them, then the FM and the business, the father and the stepmother, would not have gotten them back. Maybe it also a bit of . . . a cloud, of the uncertain future, the witch. (C4)

The conclusion of the consultant's Hansel and Gretel tale is bittersweet. For C3, the children (FM operatives) are recognized as being important:

> I think, for the first time, they get asked what it is they think FM is, what is it they think they should be doing, what is it they find challenging, what is it that they like about FM. And we do it in a day where, not only do we do stuff where we get them to talk, but we also play management games that they have never done before. (C3)

For C4, though, there is no focus on the issue of recognition. However, both tales have areas of overlap. As in the tale, the children, not only survive the attentions of the witch, but they also find their way home. This is what FM operational staff/end-users wanted throughout the tale: the household has improved, as the stepmother (senior management/main business), who instigated their perilous journey, is now sedated.

Nevertheless, the FM household is still far from the stable, loving, home of their expectations. Unlike the fairytale, Hansel and Gretel do not arrive with the treasures stolen from the witch's house. For C3, the father is about to remarry and so the cycle is likely to start again, looping continuously with the FM operative being cast back into the woods when a new partnership is formed, and with the same problems recurring again and again. C4 also observes this risk of the liminal state, but with a more optimistic prospect:

> Always the risk of a new stepmother coming in and making the same demands. Or does a new stepmum come in and the children love her. Or does a stepfather come in and completely juxtapose the conditions of the traditional nuclear family? (C4)

Here, the mere possibility of a stepfather role breaks the accepted status quo, representing a new breed of FM—one which is prepared to take a risk, break the norms of liminality, and challenge the authority figure in the relationship. In this tale, the stable, loving home is again a possibility, with all parties in the family being recognized and considered important within the overall unit. As there is no risk of being cast into the woods, there is a prospect of achieving a happy-ever-after ending, which is noticeably absent in how the consultants chose to end their tale. The 'to be continued' theme does not immediately imply an unhappy ending, but it is also far from the utopian finale that the promise of a fairytale mindset would lend itself to. This is emphasized by the fact that both storytellers have Hansel and Gretel survive, although the supernatural opponent—the crux of the tale—remains undefeated by them, leading to positioning within a 'tragic' category, which supports the findings from the *Rumpelstiltskin* deconstruction.

Overall, the story arc of *Hansel and Gretel* has clear parallels to the state of liminality, and this is mirrored in both re-imaginings of FM. There is a strong separation from what was considered the norm, as the children are cast out, into the liminal space of the woods, where the joy of the gingerbread house is matched by the threat of destruction. Hansel and Gretel enter as naïve children lured in by surface treasure but leave as perceptive survivors. 'Like the Trickersters, Hansel and Gretel enhance the paradox that spying, imitating, telling lies, and challenging property boundaries can not only enable survival but are also foundational for the discovery of how to do things with the world and to operate in an adult culture' (Tater 2015: 41).

The third and final liminal state—aggregation—is incomplete, however, in both tales: the opponent remains undefeated, and the homestead is still a place of fluidity, as opposed to a recognized and stable state. As observed earlier, FM—as a liminar—is exasperated by the short-term nature of consultancy. Nonetheless the imagery provided by *Hansel and Gretel* allowed different versions of this to be highlighted. The liminal state here is imposed on FM and not by FM itself in this case, with both consultants being aware of this perception of FM, and thus by extension disassociating their role with the accepted norm of the father role.

This is clearly illustrated by the casting of the father as FM, as:

> being scared of thinking too differently. It's almost wringing its flatcap in its hand, grateful for the work it has been given. It is very submissive . . . where it should be smarter on how it presents itself to the business. (C4)

In both re-imaginings, the FM consultants cast FM as an arm's length entity. They chose not to cast themselves—the role of FM consultant—in the tales, but rather cast internal FM or the wider genre of FM thus. The consultants in FM are comfortable in their identity, arguably indicative of the archetype of the *sage*; being brought in to help the in-house team, whose members are viewed as being unclear in their roles and at the mercy of the wider organization to shape their activities. It is this lack of control/power over functionality—and by extension, legitimacy—that the consultants cast as the malevolent influences of the witch, who remains undefeated, the *shadow*. One interpretation of why the FM consultants chose not to cast themselves in the tale suggests a wish to separate themselves from the unstable state of the FM industry, which is perceived as weak and powerless.

This initially appears at odds with the earlier theme of family, but there is a recognition that the same dangers manifested in the witch affect both wings of the family, which can only be truly defeated if FM professionals unite to empower the discipline. The relationship with management is therefore not equal, leading to a strong sense of disenchantment about FM: 'Maybe I have been unlucky in most of the organizations I have worked with, but it [FM] is there to serve rather than to advise' (C4). Both tales however are 'to be continued'. One tale, for example, predicts a landscape of unremitting change; where a new stepmother role will always be likely, and FM remains defiant against the witch. The other also recognizes this circle of mediocrity and ends similarly with themes of change and challenge, thus supporting the *tragedy* categorization. A summary of the casting to support the later ghostwritten tale is visible in Table 7.1.

Table 7.1 *Hansel and Gretel* role summary

Casting	C3 interpretation	C4 interpretation
Father	**Head of FM** Father figure in terms of hierarchical position but is not a leader. Means well but weak. *Bad Father* *Shadow*	**FM practice** Downtrodden and easily led but has a good heart. Is trying to do the best for those he is responsible for, but in a very short-sighted way. Cannot plan long term or stand up for anything important. Servant role *Material Helper/Servant* *Anima/Shadow*
Stepmother	**Senior management** Does own thing/set in ways. Outsider, command and control *Bad Mother/Witch* *Shadow*	**Main business** Source of power/does not like to be challenged. Master role *Bad Mother/Witch* *Shadow*
Hansel and Gretel	**IHFM staff** No direction and therefore lost in what they should do. Left on their own devices for survival. *Transformational Role (Winner)* *Adventurer*	**Clients—end-users (of building)** Need to depend on their own resources and rescue themselves where needs be. *Transformational Role (Loser)* *Dark Eternal Child*
Witch	**Process/Politics** Intangible process/politics always lurking in the background (and ensnaring). *Shadow*	**Fear and greed/Illusion** Manifestation of fear of change. Temptation to do the easy as opposed to right thing. *Shadow*
Typology	Tragic	Tragic

Source: Authors

7.2 The Service Providers' Storied Space: *Goldilocks*

This division of FM contains four of the ten identified stories as illustrated in Table 5.1, with the dominant tale emerging as *Goldilocks* (SP1, SP3, and SP6). This fairy story falls within the ATU category of *supernatural helpers*, and the *voyage and return* story plot (Booker 2004). Unlike the consultant category of *supernatural opponent*, here the focus is on a *supernatural helper*, and so the philosophy of the tale is considerably different. Where the FM consultants' core assumption was of a supernatural opponent that FM needed to vanquish (albeit unsuccessfully), the service providers' tale reflected no such negative forces. Rather, the category of *supernatural helpers* implies a more optimistic tale, where there is a suggestion of need for the main protagonist at the start; although this plot reflects a narrative quality of being aided as opposed to one of adversity.

The tale of *Goldilocks*, out of all the fairytales shortlisted, is one of the most simplistic in terms of characters, with only four roles being declared. Three of the four characters represent an anthropomorphic family of bears (Father, Mother, and Baby) who live together in a house in the woods. The final character is that of a young girl, Goldilocks, who enters their house and eats their food and uses their property when they are absent. Like the deconstruction tale of *Rumpelstiltskin*, this short tale contains a theme of ambiguity concerning the protagonist role, which is generally considered to be Goldilocks. The character supports the *trickster* archetype in the role of breaking accepted norms and is ambiguous morally in terms of a self-centric view (Kostera 2012).

By applying the *voyage and return* (Booker 2004) motif, the tale reflects Goldilocks's adventure into strange and dangerous territory, escaping with her life to return home a wiser and more learned individual. This is supported by the overall ATU category of *supernatural helpers*, which are the three bears. Supernatural in their ability to talk, and with other anthropomorphic qualities, they are the vehicle for Goldilocks's liminal journey. Goldilocks leaves her own world by entering their house (separation) a destructive and reckless child, and her journey through their house, using (and abusing) their property forms part of her liminal state as she tries various items and experiences to find the one that suits her: 'so Goldilocks climbed up onto the very big bed. It was too hard! Then she climbed onto the medium-sized bed. It was too soft! Then Goldilocks lay down on the tiny, little bed. It was just right! Soon she was fast asleep' (Southgate 1975: 28). Her aggregation state is illustrated by her escape from the house and therefore the bears' physical presence, which scares the protagonist but does not prove fatal. This allows Goldilocks to return to her world as an adult who has learnt valuable life lessons. Interestingly, the tale does not confirm this, focusing on the role of the three bears, but there is finality to the tale: 'by the time the three bears reached the window, Goldilocks was out of sight. The three bears never saw her again' (Southgate 1975: 8). This implies that the supernatural role of the bears—that of educating Goldilocks—is now complete.

Turning to the empirical data, SP1's lived-experience tale centred on the closure and relocation of what was considered one of the company's flagship sites. 'This site was the jewel in the crown for many, many years and it was untouchable. It was a real shock to hear that it was going to be closed and moved' (SP1). This involved the relocation of many staff to the new premises, and redundancies for those who chose not to physically relocate, as the new site was in a different county.

FM was tasked with the moving of the staff in what became increasingly tense circumstances, as many of the team chose the redundancy package. As the project progressed, the FM team diminished in numbers; but the scope of the project increased as the old estate needed to be maintained, the new site needed renovating, and temporary locations needed to be found to support the transition. Additionally, much focus began to be placed on using the change process as a catalyst for agreeing new ways of working, such as hot-desking and other shared-space initiatives. For the FM teams, investment in the new site was used as a form of incentive for staff to remain with the company, because: 'Of course you got the status of going to this new Wonderland' (SP1).

The underappreciated and largely hidden nature of FM—as illustrated in the *Hansel and Gretel* tales of the consultant—once again became evident in terms of reputational ramifications (both internal and external) brought about by the relocation process:

> It is either OK or not good enough. You might get, 'Oh what a splendid building' and 'blah, blah, blah', but you are only remembered by the first failure you are seen doing. (SP1)

The tone of the comment made here, however, is not so much one of harsh critique but more of resigned acceptance—a reflection of the acknowledged nature of the FM role in the corporate scheme of things. This position within the organizational hierarchy is thus not so much a reflection on what FM delivers, but more on how the FM role affects end-users:

> But the real stars of [the organization] are going to be the people who do the science. It will be the people who use the buildings rather than the people that make sure they are available for them. (SP1)

This would indicate a sense of humility at best, self-deprecation at worst; in the recognition that providing this service, however invisible, is accepted as part of a profession that has at its functional core the support of others.

However, the *Goldilocks* re-imagining allows a somewhat paradoxical story to emerge. As stated earlier, the tale of *Goldilocks* is essentially ambiguous; with the role of Goldilocks being reflected in a rather villainous light, one symbolic of the *trickster* archetype. This role was cast *to* the client, despite the lived-experience tale promoting a focus *on* the client, with again FM being inexorably subservient to the client function:

> If you think about it from a Total FM service, they are providing the building, the catering, somewhere to sleep and then you get someone who is coming in there and abusing them to such an extent that they are damaging some of the areas and looking for what is right for them and not necessarily what is right for the service provision. (SP1)

The three bears are cast as different levels of FM, based on the experience levels that are mapped to the hierarchy presented within the bear family structure:

> We have two developed sites, with two elderly bears, and . . . the young immature site, the young FM organization, that is desperate to please but then ends up with no porridge and comes crying to us to pick them up and help them. (SP1)

This would suggest that service provider FM is not from a world of learned knowledge, but rather from one of experience and peer help, supporting the analysis of Chapter 2 and supporting such description with an analogy of family dynamics, as supported by the consultant view of FM. There is thus a quality of naïvety reflected in the industry, as resources remain controlled and often exploited, forcing service providers to compete with each other across different portfolios.

The imagery of the bear family as FM illuminates the relationship between FM service providers and their clients. However, it can create two associations: one that reaffirms the idea of a family unit providing security and survival conditions, and another which promotes an impression of bears as predatory beings:

> In some ways, if we aren't careful, from a service provision point of view, we can be quite frightening creatures, especially when crossed. Especially when you are thinking of the customer affects—the customers' needs, and that is the chairs, the tables, your furnishings, your food, etc., the conditions in which they live. But there is the other side of that, where you are seen as some type of monster, which is what a lot of people would take a bear as. (SP1)

With the clients cast into the role of Goldilocks, and relating back to the lived-experience tale, the villainous side of the client is therefore symbolic with a possible misconception of FM. Here FM practitioners represent fearsome creatures whose professional role within organizational life remains unclear. FM service providers are aligned with the capability for causing harm—this being an implicit characteristic of their surface functionality. SP1 has a high level of awareness of this malicious character casting, which is portrayed as overshadowing what FM personnel do in creating a safe and efficient environment for the client. It is often clients themselves who are deemed the cause of destructive elements within this tale:

> The bears are FM and she is your customer. The only thing that didn't happen in *Goldilocks and the Three Bears* is that they don't savage the young girl that they found. So they were obviously there to do something other than savage her and I think that works quite well as an analogy for the FM concept. You get the bears providing the FM service and that person using them is not always happy with them. But when they find the right one, they find comfort in them. But they might break chairs, unmake all the beds, eat all the porridge. (SP1)

However, the relationship element is stressed as an overarching theme of this tale. After experiencing along with Goldilocks, the liminality of organization, the FM service provider's role is ultimately presented as one that offers accommodation and safekeeping. FM becomes the 'other' in this process, providing the role of supernatural helper by guiding the client through unsettling times and liminal states. However, where FM provides support for the client to complete a journey through liminal states, FM as the vehicle remains in the liminal state. This perpetual liminality reflects separation from the organization, and the clients that this represents. Rather than operating in unity, FM service providers remain a distinct entity—one to be treated with suspicion, for there is no clear, shared, or unified operational goal with the client. For SP1, this is illustrated in the metaphor of the two species, the human and the bear—who can communicate within their own class but not outside it:

> Goldilocks probably doesn't speak Bear—because we do speak a different language don't we? . . . So we need some common language, how do we talk in relevance back to these people? (SP1)

The role of FM service provider is therefore one of paradox; they see themselves as supernatural helpers but can be perceived by the very people they are charged to assist as a supernatural opponent. In other words, they are confined within the liminal space, with no bridging language to help relate and support what are essentially fluid organizational states. Although it can be argued that this lies within the service providers' remit to create, it nevertheless remains noticeably absent:

> I also don't think they [the clients] see the bears as service providers. But then, I don't think the bears are doing enough to help the customers in their role to help them use the service right. Maybe we need a fairy godmother to come in and act as a middleman. (SP1)

For SP1, the tale of Goldilocks carries with it a dominant theme of hope against the odds; one where service providers fashion the home for the client to prosper, despite the client not recognizing the service provider's vital role in facilitating this state of well-being:

> You don't want your customer running off at the end of that story do you? You want them living in harmony, and growing up to love them as cute cuddly bears, and see them for what they are—good porridge makers and soft bed providers, and neat builders of chairs. So I would have it more as a happy ending if I could. Although there are some customers that I think could do with tearing apart! (SP1)

SP6's lived-experience story had a similar plot line to SP1 with its focus on property: a 48-hour building shutdown, completed for statutory purposes, and affecting twelve thousand occupants. This tale concentrated on the technical aspects of the project, reflecting the 'hard' FM background of the participant. Aligning with SP1, the organizational condition of FM was again seen as misunderstood. In contrast, however, the construed role of the client vis-à-vis FM is not so much seen as that of a relationship but of a preferred separation: 'If you start going into great complexity of what we're going to do I think they either lose interest or just not understand' (SP6). Both tales reflect that this is an accepted role within the storied space of FM service providers:

> People didn't understand the complexity behind it and how big of an actual achievement it was . . . because I think that companies potentially see it as sort of BAU [business as usual], you know, that's what we're here to do, that's what we're here to be paid for. (SP6)

In this tale, the nature of recognition for FM service providers came from winning and renewing contracts as opposed to successfully undertaking individual actions: 'You're not going to pay someone millions of pounds if you think they're crap' (SP6). SP6 found recognition by looking internally within the service provider industry rather than forming an identity through the perception of the client (McCarroll 2017). SP6's storied space therefore recognizes the power within his job role, focusing on the relationship with the building rather than the organizational actors within it:

> To see a forty-storey building completely empty, no one allowed into it, all the lights cut, because there was no electric being fed, with them all sat outside and just looking at this huge building in pitch darkness. And it was that control that we can have on a building, how much it relies on us being there. Because if the electric went, it just showed you access control systems can fail, lights fail, computers fail, and it just showed that if you take one element of FM out of that building, then everything shuts down. (SP6)

As with SP1, the client is cast in the role of Goldilocks. But in this re-imaging, there is much less ambiguity over the status of the protagonist. In this tale, Goldilocks is not naïve in her actions within the bears' house but is accomplishing them purposefully to ensure the best product/service is provided. This is not a malicious act, but one that is intentional and in the best interests of the character:

> Goldilocks would be more of the client, because you know the clients go out and have a taster of what FM providers are out there and see what suits them for their particular needs. (SP6)

However, the negative impact of this is seen in the casting of the bears as FM. The three characters are interpreted as a collective, and although the lived-experience tale is one of power, the imagery ascribed within the fairytale reverts to the association of FM and janitorial duties. This illustrates that such imagery is dominant in relation to the service provider sphere both externally and internally, supporting the analysis of Chapter 2. Here, it is the damage caused by the protagonist that SP6 is left to resolve, and it is this arrangement that led to a casting of roles in a similar manner to that of SP1:

> If someone comes in and makes a mess in the toilet and runs away from it, and you've got cleaners coming in to clean it up, and plumbers coming in to fix it, and the people that have done it have just ran off and thought nothing more of it. (SP6)

This therefore positions the client within the (negative) role of the *eternal child*, avoiding responsibility and commitment (Kostera 2012). The imagery of the bears as representing the perceived villainous nature of FM is also visible within this re-imagining, but here there are more positive connotations of perception:

> They [the client] want to have their photos, and their bottles, and the pen holders all around them, and cleaning teams are saying, no, you can't do that, you're the villain, but you've other people that are moving with the times and want to move around and want that clean desk and that they get it and they understand, so it's a mixture of both worlds. (SP6)

The last service provider story comes from FM procurement. The central narrative here is SP3's assertion that her 'primary role is to win work'. Her tale centred on working with operational FM teams and suppliers in order to create innovative efficiencies, such as building cost/time savings into tender bids and then presenting the latter to the FM board. For SP3, the role of procurement is a vital part of service provider FM: '*We are the conduit aren't we, we are the relationship*' (SP3). This functional emphasis on the 'relationship' was also evident in the account of SP1 but is notably absent in that of SP6.

However, in SP3's lived-experience story, this emphasis on relationship management, and notably the role of innovation within it, is not shared by the board:

> I don't know whether it was funny, sad, or it was just very interesting that, you know, I think the general conversation around the table was, yes, we should be doing this, but nobody had any idea what the next steps should be or how we could take that forward. I asked for buy-in from the Operations Team and there was more or less complete silence around the table . . . there's a lot of kind of, we did this three years ago, you know, and nothing came of it, and you know, you're almost kind of wasting your time. (SP3)

SP3's tale revolved around attempting to get support from her colleagues and her frustration at being unable to do so. This was attributed to service provider FM being outmoded in its approach, and a perception of general torpor in the industry, which lead to several rifts ultimately affecting delivery:

> It's a bit set in the dark ages, FM still. So people like me, who have only been here for two and a half years, I'm sort of still quite enthusiastic and trying to change things and kind of drive things. You can see once you've reached a certain point, you know, once you've been here for *x* amount of years, you then turn in to the glass half-empty, you know, we've tried it before, loads of blame . . . you really do get that kind of . . . support function divide. (SP3)

This tale is internally focused, with little reference directly to the clients, but the importance of the relationship is still recognized by SP3. This is a tale of frustration and tragedy that reflects the internal divides within service provider FM. However, the casting of Goldilocks reveals a more positive projection for the future. Where SP1 and SP6 cast Goldilocks as the *trickster* archetype client, here this role is allocated to service provider FM itself, as directly experienced in the career of SP3:

> I've joined this industry and joined thinking, God, there's so much opportunity here, it's not the finished article, hmm, there's loads of things I can do and over the period of time I'm / I still keep trying the porridge, if you see what I mean, I still keep dipping my toes into things and doing things. But I'm almost getting to the point now where I wonder how big the appetite is for change. Because there's almost an element of if it's not broken just don't try and fix it. (SP3)

Here, SP3 reflects a personal journey that is liminal. The discipline of FM was perceived as large and bountiful, with lots of different options and opportunities to explore, here linking the role of Goldilocks with the *adventurer* archetype. However, once within the industry, the options were perceived as less accessible in what was perceived as a more static environment. Still seeking the 'just right' element that FM as an industry promised, and yet did not deliver, has led to this story also being one of frustration, although not directed so much at the client base, as with the other two retellings, but instead focused inwardly, to the industry itself. The three bears here are FM colleagues who operate as decision-makers, signally holding a position of power indicative of the negative *king* archetype. The bears are the housemakers who wish to maintain the status quo, reneging on their wider responsibilities, and being the main force inhibiting innovation. For SP3, this lack of drive and change within service provider FM is likely to end in her withdrawal from FM:

> It gives me a personal dilemma I guess, as well, because if there's no appetite to change, and I really can't change anything, probably I'm not going to be personally satisfied staying here . . . if I'm not able to bring change . . . I will be skipping off into the woods. (SP3)

The retelling of *Goldilocks* as chosen and narrated by service provider facilities managers reflects the tale type of *supernatural helpers*, but in each case supernatural help was provided by FM rather than FM receiving the help.

For SP1, this was perceived as the natural order; FM's role was to be subservient for the greater good of the organization it was contracted to support. For SP6, supernatural help is delivered to the unwitting client and illustrated in their power to shut down all utilities in large office buildings. For the last tale, supernatural help was required but was not forthcoming. This is manifested in the desire of FM service providers to help other colleagues in initiating change that would lead to their work unit's own long-term survival, a circumstance intertwined metaphorically with the client's own perceived fate, as illustrated in the *Goldilocks* story. All three tale re-imaginings have suggested that FM was not recognized for the work it completed, or the functional potential it holds; with the first two proposing this is a view held by the client and the last that this was also held internally within FM. The different interpretations witnessed across the three interviews are mapped in Table 7.2.

All three re-imaginings therefore possess the story arc of *tragedy* or *tragic-epics*, SP1 sees their role as dealing with the *trickster*, with the aim to educate them for a united better future. This is an *epic* tale, which currently ends in tragedy, as they fail to achieve this end goal. SP6 also fails to educate the client, who aligns with the negative qualities of the *eternal child* archetype. SP3 also fails in their happy ever after with the evil *king* of FM preventing their *adventurer* archetype from completing their quest.

Table 7.2 *Goldilocks* summary

Casting	SP1	SP6	SP3
Goldilocks	**Customer** Is reckless but naïve, does not see the difference in good or bad, breaks accepted norms. *Dark Daughter/ Dark Tricksters*	**Customer** Is disinterested or incapable of understanding/ taking responsibility, concentrated only on self. *Dark Daughter/ Dark Eternal Child*	**SPFM procurement** Trying to find the best way in a new environment. *Adventurer*
Father Bear Mother Bear Baby Bear	**Experienced/Novice FM manager** Provider of shelter and security, but feared as predator. *Light Father/ Sage*	**FM** Provider of shelter but power is not recognized or appreciated. *Light Father/ Sage*	**FM** Rules without empathy, enslaves followers, inability to lead/change. *Dark King*
Typology	Tragi-Epic	Tragi-Epic	Tragedy

Source: Authors

7.3 The In-house Facility Managers' Storied Spaces: *Cinderella*

This division of FM represents the largest section of the participants, although only three of the ten available fairytales were selected by this group, as illustrated in Table 4.6. This category was clearly dominated by the tale of *Cinderella*, with five of the eight participants identifying with it (IH2, IH4, IH6, IH7, and IH8). This supports the FM industry associating itself with being a Cinderella industry as discussed in Chapter 2, and Section 6.5. Like the service providers, the ATU categorization was that of *supernatural helpers*, although this tale is then matched to a *rags to riches* typology (Booker 2004) which is suggestive of recognized and celebrated success.

IH2's lived-experience story is one of change, where the FM property teams that dealt with residential and commercial estate management merging into one department. Concentrating on mainly the statutory maintenance of these properties, the main story involved the transition period, concentrating on the dislike the commercial team felt towards the residential team now managing some of what they considered 'their' estate:

> They didn't communicate with us very well and we tried to communicate with them but always found ourselves fighting for information and it always felt as though you were in conflict with them when going to meet them. They were always trying to pick you up on something, no matter how trivial. (IH2)

This aligns with the lived-experience story of IH7, which also dealt with the merging of two existing FM departments to form one united one, driven by external forces of a wider organizational reorganization. In the case of IH7, it was small soft services department that was merging into a well-established hard FM division. Both tales relayed a lot of internal tension, with IH7 stating that the hard FM division were:

> almost like the gods, the kings, sometimes we referred them to as Dad at work and what they said went basically, in terms of statutory compliance . . . the Cinderella, you know, all the soft services, it's sort of like evolved from the caretaker role and it was seen very much asa soft aspect of the delivery model and the engineers sort of referred to the cleaning, it's just cleaning . . . and it had no stature, no boundaries and it wasn't defined as facilities management. (IH7)

The lived-experience tale detailed the internal conflict of the new in-house team, with little acknowledgement of the client base in both IH2 and IH7.

The frustration held by IH2 was evident, and that it caused the team to remain static, fighting to provide the same service as before, rather than any advancement, which was the original purpose of the reorganization: 'It was purely fighting to stand still' (IH2).

This was echoed in IH7, although the merger here was supported by a change in the FM personnel in the hard FM division, which eased the tension:

> The engineering type of role is a dying breed, so it was a certain number of people, particularly men, of a certain age, that sort of carried the stature within the Estates Department and yes, they sort of kind of retired or they had a restructure and made a lot of redundancies. (IH7)

Both IH2 and IH7 focus on the re-imagining element of the interview, with the lived-experience stories being brief in comparison. For IH2, it reflects the storyteller's personal career journey in FM, whilst IH7 concentrates on the FM department. Both have in-house FM positioned as the main protagonist.

For IH2, Cinderella is part of a happy, functional family at the start of the tale, until the death of her mother, which is associated to the death of the old residential department and the introduction of new members: the commercial stepmother and sisters:

> Cinderella, the assistant FM, is/has more work to do, has the budget taken away from them, can't deliver the services that they used to deliver to the company, so she seems to be really unhappy. (IH2)

Demoted in the new structure, Cinderella/IH2 sends out her CV, but her details get lost. She later meets the recruitment agent again by chance, and upon producing her CV, the agent recognizes that she was the candidate he was searching for. In this way, IH2 casts the supernatural helper role, the fairy godmother, as the recruitment agent. The happy ever after is provided by the new role, the prince, representing the balance of animus and anima. The king is uncast.

IH2 achieved his happy ending in this tale, but only by empowering the protagonist to leave the current FM organization and use her skills to secure a better match:

> And so it was that Cinderella left the business and started a new job working for a company that ensured she had the resources and budget she needed to support the business. (IH2)

This would indicate a toxic element within in-house FM, and the only way IH2 could see a feasible future was to seek a new role, although he chooses to remain within the kingdom of FM. The supernatural helper appeared in the form of the recruitment agent, although it was Cinderella's true qualities that made this encounter a success.

In contrast, IH7's Cinderella tale revealed rich details of the merger, which was a successful transition in line with the *rags to riches* story arc:

> The Cinderella story was most akin to what I relate to in terms of where we've come from and where we are now, on the basis that it's very regal and royal, and I can associate that with the old hierarchy. (IH7)

In contrast to IH2, this is a story of internal support and change together, as illustrated by the casting. With Cinderella as the in-house FM, the domineering and male status of the hard FM division from the lived-experience tale led to their casting as the father/king roles interchangeably. 'It was male dominated, there was no . . . female representation in the estates department and they were sort of seen as like the father, the king' (IH7). Not only was gender a leading factor in the casting here, but it was also the legacy of the status of the hard FM department, who held the patriarch figure and which only changed as they collectively retired and exited from the tale.

The villainous character within the tale of *Cinderella*, that of the stepsisters, revealed the antagonistic relationship held with the client, who in this case were clinicians. This was illustrated in the lack of voice that FM had in the face of the stepsisters/clients who did not recognize their sister as a valuable member of the family whose fates were intertwined:

> Estates always sort of got lost out of the decision-making process and clinical provision always won over the estates need, but what they didn't realize is that, you know, the estates was the enabler to their delivery and sometimes that got forgotten. (IH7)

Turning to the supernatural helper roles of the fairy godmother and the prince respectively, these were cast as external FM entities, that of consultant and service provider FM. The consultant FM:

> managed the in-house bid for cleaning services and that was a major turning point for FM, because . . . it suddenly was brought to life the amount of money that we spend on cleaning, the specification and the logistics and the delivery, people started to respect that, hang on, it's not just cleaning, it's multi-faceted, it's difficult, even resourcing such a service is difficult, etc. and because it had / because of

> the size of the tender bid, it sort of pulled on all other functions, finance, HR, and people started to then sort of see what FM is about and it exposed it a little bit. (IH7)

The supernatural power lies in the influence of the external pathway of FM, who were able to raise the profile of FM by exposing the cost to the overall organization and reveal the power of FM. This was their magical ability that in-house FM were powerless to do and needed help to unite all the different internal factions of the organization.

This magical ability of external FM is carried through the service provider function, who are another valuable ally, playing a large role in the success of the in-house FM protagonist by providing other skills that were absent in-house:

> If you get a good outsource provider, they are almost like the prince, because in FM there is no technical capabilities internally, so you are relying on that outsource provider to not just deliver a job, but to support you. (IH7)

The supportive roles of the coach and footman are cast as front-line, operational FM. In this case, being the porters who play an important and yet fairly invisible role within the overall story:

> The coachman and the rat, the footman and lizard, always reminds me of the portering services and the mail room, I don't know why, I think it's just because it's that / it's that movement isn't it, that logistical approach, as well as the soft nature you'd expect from the footmen . . . they're almost like a caring / like a caring big brother sort of character. (IH7)

In IH7, this tale is one of a happy ever after, an *epic* tale where the newly formed kingdom of the merger has formed secure partnerships with the service provider divisions of FM.

The lived-experience story of IH4 was also one concerning an organizational merger—where two client bases came together and thus the in-house FM departments followed suit:

> Two banks had merged together, and they had two FMs, and what I didn't know of in the beginning, it wasn't explained to me, that there was some animosity between facilities managers . . . there wasn't a great deal of due diligence done as far as support services... They were both jockeying for position. (IH4)

The result of this was that one of the main facilities managers had taken long-term stress leave, and IH4 was covering that role. The remaining FM manager was now responsible for the whole portfolio and reacted badly to IH4's appointment: 'she was trying to just put walls up etc. and she didn't want to let me in' (IH4).

In response, IH4'S tale was one of building a new department, relating back to the change stories of the consultants and service providers. He started to talk to stakeholders, develop new policies, and complete some visible work such as refurbishing some areas to encourage the two different divisions to work together under their new identity as equals: 'A little bit of decoration, a few plants here and there, it really changed people's outlook and the fact that, you know, they weren't the poor relation as such' (IH4).

This proactive approach was new to the established in-house teams, both in terms of their internal service and that of the merger they were asked to support. The main in-house FM began to accept IH4's presence and impact, becoming less hostile and more engaged:

> She'd been in the building a long time, maybe, I don't know how many years, a few weeks or a couple of months, you know, she hadn't really spoken to anybody and so I mean how difficult it was for her I'm not too sure, but yes, I certainly / I know for a fact it helped her and I know I made a difference there. (IH4)

The ending of IH4's tale ended when he was offered another role in another company: 'what happened after that I'm not too sure' (IH4). His story is one of being the saviour of an ill-prepared and ill-supported FM department, which was unable to change themselves never mind support the wider organizational change that was underway:

> I think that the biggest thing was the building on the development of the relationships, but at the same time, just to go back, I just think that a lot of due diligence could have been done with support services . . . I just think it was completely missed and I don't think they realized the importance of facilities management. (IH4)

Unlike the stories analysed to date, the re-imagining of IH4's fairytale was not based on the lived-experience tale. However, a similar story arc was presented, with FM as the hidden heroes of the organizational kingdom. In this tale, FM assumes the role of the main protagonist, and is specifically not the in-house storyteller himself, but that of a member of his staff who was a service provider FM employed as a security guard:

> He's been through several contract changes . . . and when I first met him he was quite shy and quite retiring . . . although he went above and beyond, he just didn't

> really have any belief. He didn't have any belief in himself and that's Cinderella in a way . . . He's just a really, really good person, like he's been through several . . . acquisitions and several changes of contractor. (IH4)

In contrast to IH7, where the supernatural help came in the form of external elements of FM, such as the service providers and consultants, here, they are the subject of rescue, positioning IH4 in the role of the fairy godmother, which came as some surprise to the in-house storyteller himself:

> I'm the fairy godmother . . . I don't believe it . . . Oh my God! How crazy is that! . . . when I realized what he was doing, I made sure that he got a substantial pay rise, what he was actually doing for us. I included him in a lot more stuff and he wanted to develop, he wanted to do more facilities stuff, and although he is still a security guard working for [service provider FM], I include him in a lot more of what's going on. (IH4)

The power represented by the magic of a supernatural helper lay within in-house FM to help the protagonist, the service provider FM security guard, to receive the help (training) to evolve to a different identity, recognizing the inherent skills and potential within the character and ultimately recognizing him as a valuable part of the family. The aligned role of the supernatural helper, that of the prince, was interpreted as receiving better working conditions, which allowed Cinderella to showcase her skill base in a supportive environment—where she was meant to be and recognized as such. The king/father roles were uncast:

> Now he knows who he's dealing with and he's getting confident and he's willing to speak up etc., whereas when I first met him it's not the type of thing he would have done. He just feels included. Silly little things like / maybe not silly, but he's actually / he comes to our Christmas party and we actually pay his company for another guard so he can come to the Christmas party . . . so although he does work for a third party, he is more like / I treat him like one of our own, I don't know whether that's right, wrong, good or bad . . . he's part of the family. (IH4)

As with IH7, the stepsisters are cast to the role of procurement. This does not reflect a villainous nature but one rather that does not recognize Cinderella and therefore excludes her without thought: 'the stepsisters were the people that . . . just change security, you do that, you do that, you're on time, and duh duh duh, and was / there was no real love there I guess' (IH4).

Overall, IH4 was very articulate when exploring the fairytale and its rags to riches narrative in relation to his own personal journey within the in-house FM profession:

> I guess it's a reflection on my own story, because I just started off like in an office, just as a facilities officer ... and now I'm you know... fingers crossed I'll be director of facilities in April, be promised a directorship. (IH4)

This theme of helping people in in-house FM is reflected in IH6, whose lived-experience story reflected several incidents of the in-house FM team pulling together to meet the unrealistic deadlines of end clients, such as delivering a late catering order, and providing out-of-hours services:

> I really like the people that I work with and the fact that you're involved in so many people's lives, by delivering a service . . . sometimes the first people they might see aren't the ones you'd want them to, so people arrive at weekends, they arrive out-of-hours, it could be a porter, who has been there for thirty years and is a bit rough and ready, but delivers a great service. (IH6)

In contrast, IH8's lived-experience tale is also one of being called on to deliver change, but reflects a tale of being a lone voice, with no recognizable in-house team to call upon. Charged with introducing an open-plan policy, the end clients who were the focus of this project reacted very badly within the initial consultation groups that IH8 had organized. 'People actually showed physical anger, so they would bang the desk with their fist, and they weren't accusing me, but they were accusing [head of division] of bullying them out of their offices' (IH8). Here, it is FM that are positioned as having to deliver a very unpopular scheme, having to lead negotiations over a decision that they had no voice in making, but all the responsibility to implement. In this lived-experience tale, in-house FM was portrayed as both the villainous entity to the end clients, but also as a powerless entity for higher management.

IH8 like IH4 is a tale of resilience over resistance, where IH8 tried to manage both the end-users' expectations but also those of higher management. The frustration of the situation was paramount in this tale, where both parties became increasingly aggravated: 'I can't get them to listen and open their minds; it is a change-management issue' (IH8). This story concluded when senior management withdrew support, and the project was abandoned. In-house FM therefore remained the villain, and the real aim of the project to improve working conditions was lost. It was this lost message, and the opportunity to change for the better that IH8 lamented the most in this tragic story. It was in the:

> name of efficiency, rather than effectiveness, and that probably makes me feel quite sad because it makes me think actually that wasn't what we were trying to get across in the early days, we were trying to say, not how many desks can be squeezed into this space, it's how might we work differently across subject groups, how might we work differently in our office. (IH8)

The lack of recognition for successful works, as seen in previous stories, is once again apparent in IH6 and IH7, aggravated by scrutiny of perceived errors:

> I think there is that lack of recognition sometimes because if you're doing a good job people don't know you're there, hopefully, and the first thing goes wrong then obviously everybody is upset because they've got no heating, water, it's not clean, so yes. (IH6)

However, this lack of recognition on works completed is also illustrated in the lost potential of FM. Both IH6 and IH8 illustrate the aim of in-house FM to be cast as an important ally that can actively improve everyone's working conditions, being led by a concern for the people themselves. Where IH8 feels that the project failed due to this lack of understanding of what FM was trying to deliver to the end-users, IH6 turns this principle of helping people back towards the in-house FM team, feeling it is important to create 'an ethos that if there's something not happening right, everybody is willing to join in' (IH6).

Turning to the re-imagining of *Cinderella*, both IH6 and IH8 display an internal focus, both casting themselves within the protagonist role. The imagery of being invisible, despite actively contributing to the household, is once again evident:

> We were Cinderella, which is what you have there because she is sort of working away, sometimes it seems thanklessly (laughs), while potentially there are sort of stepsisters out there taking the glory on her behalf. Yes, so I kind of thought it was that—sort of behind the scenes, completing fairly thankless mundane tasks. (IH6)

Where IH8 casts her protagonist as helpless in this situation, IH6 shows element of the riches indicated in the story: 'every now and again that there is a bit of recognition and potentially does get to go to the ball' (IH6). This is still an irregular occurrence, but illustrates that these incidents not only do transpire, but are also recognized as important within the FM landscape. This echoes the sentiment of IH4, who emphasizes the importance of inviting the security guard/Cinderella figure to their Christmas party, showing that the ball is not only a sign of recognition but also inclusivity, of creating a sense of family (IH4).

Turning to the fairytale, the traditional head of the family, the father, holds a shadow position in this tale. As such, he is uncast in IH6's tale—reflecting a more equalitarian approach—but more prominent in IH8's—where as head of FM he fails to utilize his role of perceived protector:

> He was, you know, really passionate about trying to understand how the different ways of working might work . . . and I do think he genuinely thought about that, but he was actually / he was too scared to ever do anything about it. (IH8)

The majority of in-house FM tales left this role uncast, with only IH2 and IH7 allocating this role to redundant figures—directors displaced by mergers. In IH8, this role echoes a source of power that is not released, and in alignment to the fairytale, there is a sense of abandonment from an authority figure who was meant to protect.

The disconnect within a perceived family unit also follows through to IH8's casting of the stepsisters. Following the pattern across most stories, this slightly villainous role is cast to the client. For IH8, they are the working group that formed part of the open-plan consultation but were only interested in their own areas as opposed to the wider plan. As the sisters not only abandoned Cinderella, but they actively were also malevolent to the protagonist, which IH8 identified strongly with:

> A sense of people could have stepped up to the mark to be more supportive of the consultation exercise and if they were sometimes / they were present within the consultation sessions and other people who were in there they could have also stepped up when people were being rude, slightly defamatory . . . I guess what I'm looking at is . . . that these are meant to be my sisters, these are meant to be people that will support me. (IH8)

This distance is echoed in IH6, where the client as the stepsisters is not proactive in their negative attitude to the protagonist but are actively separate from in-house FM: 'they are up there and they're in their ivory towers' (IH6). It is within this setting where the *rags to riches* conclusion can only be completed with the aid of a supernatural helper, the fairy godmother. With the father (head of FM) being perceived to abandon the project for his own self-interest, the role of the fairy godmother is emphasized for IH8. Cast as the head of department, who initiated the open-plan office project, she withdrew her help and support in the face of hostile feedback. In this way, IH8 viewed the supernatural helper as one who wanted to be viewed by all parties as a loved figure but not one who was matched to help in-house FM. This character then was the supernatural helper to the end-users, to the active detriment to IH8's ability to deliver the project that ironically, the fairy godmother had begun:

> She wants to be seen as the person that would listen to the stories. She was receiving the results from the consultation exercise and she was saying I will not do this to you; I will not impose this change on you. She was gaining a positive reputation through not driving the project forwards, through making sure that the rat and the lizard didn't turn in to the coachman and the footman. (IH8)

IH8's story, therefore, is one of an isolated protagonist, with this motif being linked to the absence of a protector evident in IH6's account. For IH6, however, this manifests itself internally. The source of the help lies within the in-house FM team and their ability to look after each other and develop a recognized legacy:

> We might act as fairy godmothers, by trying to raise the profile of some of these staff and getting them recognized for external awards or trying to develop them and roll out things like accredited training. (IH6)

It is through the spirit of unity and support that IH6's Cinderella story becomes one of internal empowerment, with the *rags to riches* story being illustrated through staff, who have been supported to realize their own skills and therefore rise in the hierarchical ranks. The prince here is the enabler of this network, who provides the opportunity and resources for those he is responsible for as a member of his kingdom and their continued presence:

> I'm not by any means saying our director of FM is a prince, but I guess if they were going, 'Do you know what, there's no money for development, and we're not going to invest in people,' then the whole thing would fall and Cinderella would just be scrubbing and there would be no chance of moving up or moving out or going to the ball or whatever, so yes, I guess there's got to be somewhere higher up the chain that will allow managers to manage, managers to develop, managers to reward, there's got to be an appetite for that somewhere I think. (IH6)

It is this alternative ending that lies within the Cinderella story of IH8. With no support and open adversaries, Cinderella does not make it to the ball, and there is no prince to meet. However, this is not the end of IH8's tale, and the resilient nature of in-house FM is once again apparent. Here the rags of the story are fully apparent, but the time is not right for FM:

> You've just got to, you know, live with the fact that you're not going to be a fairy godmother to these people . . . A bit of *Sleeping Beauty* in there because actually, in the end . . . maybe Cinderella's being poisoned and gone to sleep, umm, yes, maybe she's been poisoned and gone to sleep and actually when the time's right she'll be woken up and she'll live to fight another day. (IH8)

As an extended metaphor of the storied spaces of in-house FM, the five re-imaginings of *Cinderella* reveal an essentially internal focus, with a small number of roles being cast throughout the tale (as illustrated in Table 7.3).

Three of the five lived-experience tales detailed stories of FM mergers that were created in response to wider organizational changes (IH2, IH4, and IH7). This highlighted internal divides that proved to be detrimental to FM in general, although it should be noted that many of these stories ended positively.

All but one cast in-house FM as Cinderella, the protagonist in need of supernatural help, which comes in the form of colleagues within the in-house FM team, meaning that there is little expectation of external help, in terms of either support or resources. However, IH4 and IH7 are notable exceptions here. In IH4, it is the in-house provider that is the empowered party, who comes to the help of their service provider Cinderella, recognizing his skills and making him part of the family. In IH7, it is external FM, in the form of consultants and service providers within FM, who bring the necessary skills and knowledge necessary to survival to the in-house provision. They are united under an FM umbrella.

Across the board, there is no help or recognition from the organization FM supports. Indeed, the client was notably absent for the majority of the lived-experience tales, considering the service delivery nature of this secondary service, but appeared in all but one of the re-imaginings. Here they were universally portrayed within the villainous role of the stepsisters, either fully malevolent or through ignorance of the shared benefits a more united approach could result in. The recognition that the supernatural helper role is not from any organizational entity outside the world of FM emphasized the separation of this discipline from the corporate entity it is charged to support,

Table 7.3 *Cinderella* summary

Casting	IH2	IH4	IH6	IH7	IH8
Cinderella	**IHFM** Victim of circumstance, pulls on own talent. *Anima/* *Good daughter*	**SPFM (Security guard)** Shy but talented, overlooked. *Anima/* *Good daughter*	**IHFM** Completes dirty jobs but goes to ball sometimes. *Anima/* *Good daughter*	**IHFM Soft Services** Talented underdog undergoing change. *Anima/* *Good daughter*	**IHFM** Underserving victim trying to appease conflicting interests. *Anima/* *Good daughter*
Father	**Old FM teams** Fear change, power centric. *Animus* *Dark King*	Uncast		**IHFM Hard Services** Out of touch, retiring male FM dept. *Animus* *Dark King*	**Head of IHFM** Means well but feeble and ineffective. *Animus* *Dark King*
Step/Ugly Sisters	Uncast	**Client procurement team** Powerful and uncaring. *Animus* *Dark Daughter*	**Client** Powerful and uncaring. *Animus* *Dark Daughter*	**Client** Not recognize FM as same family, elite and powerful. *Dark Persona/ Daughter*	**Client** Work completed in self-interest. *Dark Trickster/* *Daughter*

King	Uncast		**IHFM Director**	See Father	Uncast
Prince	**New FM Role** Recognition of talent. *Balance of animus and anima*	**Better working conditions** Recognition of talent. *Balance of animus and anima*	Developing role, needs support from followers. *King*	**SPFM/ Consultant FM** Exposes hidden truths, Brings new resources and a powerful reputation. *Spiritual Helper/Wiseman Sage*	
Fairy God-mother	**Recruitment consultant** Provides escape route to new better environment. *Wiseman/ Sage*	**IHFM** Recognized talent and enabled Cinderella to recognize her talents. *Wiseman/ Sage*	**IHFM** See Cinderella—support system. *Wiseman/ Sage*		**H.O.D** Led vision but withdrew support when proved unpopular—wanted to be loved figure. *Dark Persona/ False Prophet*
Typology	Epic (Rags to Riches)	Epic (Rags to Riches)	Epic (Rags to Riches)—journey not complete	Epic (Rags to Riches)	Tragedy

Source: Authors

and in the case of the in-house FM, are actively a part of. With no supernatural helper within the main organization, there is no guidance on how to negotiate an accepted entry, of how to be *invited in* (Harris et al. 2013) to achieve the aggregation closure state of liminality.

7.4 Coda

This analysis has illustrated how each of the three FM divisions possesses its own storied spaces (research aim one) and archetypes (research aim two). It illustrates not only how divisions create their own stories and accounts, but also how they view those of each other. The factual stories represent the overt and literal analytical layers—a relaying of facts—whereas the fairytales capture the tacit, contextual, and emotional realities experienced by FM practitioners (see Kociatkiewicz and Kostera 2012).

The FM consultant analysis revealed an overarching storied space of an enemy that needed to be overcome. Using the ATU categorization, it highlighted the role of a *supernatural opponent* and told of FM being at the mercy of an insuperable entity. In terms of archetypes, the deconstruction analysis cast the consultant in the role of the miller's daughter, tasked with the impossible and finding the answer through luck. This lack of influence in the face of an all-powerful opponent is also present in the construction tales, where the consultants cast FM as the father of Hansel and Gretel. These tales present the consultants' view of FM as manifesting itself as the 'position of serfs, expected to be purely an item within the narrative of their master or mistress' (Sims 2003: 1196). This results in highlighting the *dark father* and *material helper/servant* archetypes (Kostera 2012), reflecting forces acting to the detriment of the discipline. This may explain why the consultants did not cast themselves directly; thus, creating distance between their role and the perceived discipline of FM—portraying an interpretation that FM consultants wish to be positioned outside the FM story itself.

Meanwhile, the FM service provider grouping represents the most varied in terms of categories and tales. Whereas their consultant counterparts align with the storied space of *supernatural opponents* who remain tragically undefeated, FM service providers align instead to ATU tales of *supernatural helpers* and *voyage and return*. These tales represent FM service providers as both altruistic givers of help (e.g. the bears in *Goldilocks*) and underserved receivers (e.g. *Rumpelstiltskin*'s miller). In both the deconstruction and construction and tales, the FM service provider is not so much cast as the main protagonist, but rather (and similar to the consultant's tale) as a member of the essentially powerless supporting cast. It is, for example, not necessarily the FM service provider's voyage that is the focus of concern, but rather their role—as a *sage* archetype—in instigating and/or supporting the voyages of others. These tales are unique in that they illustrate not only the more malevolent power of FM—which can be used for exploitation—but also FM service providers' own awareness of this quality in how they are beheld, especially from a client perspective.

And turning to the accounts of in-house FM, the chosen tale of *Cinderella* again reflects the ATU category of *supernatural helper*—although here reflecting a successful *rags to riches* story arc rather than *voyage and return* (Booker 2004). In contrast to the tales of the consultant and the service providers, the in-house FM story focus has a strong internal concentration but a wider acknowledgement of their service provider and consultant counterparts. Where the deconstruction carried with it a focus on internal politics, the construction tales supported this with the lived-experience tales reflecting the turmoil of enforced mergers, both in terms of the FM teams, but also in their role of supporting wider organizational change.

In-house FM sees its role once again as the *sage*, successfully helping the client survive in a liminar between the two worlds of senior management and end-user. While the liminal space metaphor is discernible in all three FM divisions, the only illustration of a positive story arc in this regard is in the re-imagining accounts of in-house FM. Unlike the service provision tales, which omitted the role of FM consultants, the latter are cast by in-house FM as allies that can help with their quest. This contrasts markedly, however, with accounts of in-house FM's own senior management team, which is perceived as dishonest and deceiving. Similarly, senior managerial positions in the organization are cast as villainous, despite the fact that this is not a separate organization, as would be the case for consultants and service providers.

In sum, the deconstructed and constructed accounts revealed three distinct story spaces and archetypes. First, the consultant tales were wholly *tragic*, exhibiting dark story arcs and weak and unstable archetypes. It is notable that the consultants distanced themselves from any direct casting, preferring to cast FM as a discipline rather than as the role of a specific consulting individual. This suggests awareness of perceptions of FM as a 'veneer' (Varcoe and Hinks 2014a)—an identity they may not wish to identify with. Second, when developing their accounts, the FM service providers shared with the consultant group a principal recourse to the *tragic* typology. In so doing, however, neither service providers nor consultants referred to one another when making interpretations. Instead, they referred only to their in-house counterparts, who were viewed, in the main, as an overlooked entity—an initially naïve grouping that is nonetheless capable of becoming their own hero. And third, and conversely, accounts by the in-house group referenced both of their FM counterparts, consultants, and service providers. In so doing, such references were generally complementary—describing FM professionals who were left without adequate operational direction yet who still delivered, in the process bringing valuable resources from the 'outside' while exposing internal faults from their position in liminal space. The in-house tales stand alone in developing a more positive storied space, one that makes recourse to elements of *epic* and *rag to riches* categories. A common factor across all three groups, however, is the negative casting of the client, while, similarly, negative archetypes are aligned with senior management, whether representing FM or the organization as a whole.

8
The *Artificial Folklore* of FM

Building on the analysis of Chapter 7, this chapter turns to the interpretation of the data, highlighting various patterns within FM's storied space. In the process, four key findings are identified in support of research aims one and two. Throughout this research, the role and character of fairytales was developed to frame the storied spaces of FM. In so doing, stories were defined, and fictional narratives drawn upon, in order to identify symbolic meaning through the interpretive research process (Weick 1995). As a result, this chapter culminates with a ghostwritten fairy story from each of the three divisions, completing the *artificial folklore* approach outlined in research aim three.

8.1 Key Finding One: Issues of Lack and Organizational Illegitimacy

All three divisions of FM—consultants, service providers, and in-house operatives—chose different story categories and tales within the fairytale framework. On the surface, this would suggest the absence of a consistent story typology within the industry, with few tales being shared across practitioner groupings. However, what arguably unites the discipline in this regard is that which is not immediately apparent: a shared acceptance of lack (Lacan 1977). As analysed in Chapter 3, fairytales are associated with a utopian ideal, but the reality for FM is the absence of good, and therefore dominance of the *shadow* archetype. This suggests villainy or the desire to have something that creates a journey the protagonist must complete.

For the consultants, this absence of good is related profoundly to their journey typology, identifying a *supernatural opponent* that tragically is never overcome. They are a woodsman cast into a father role, but this is a role they have no resources or skillset to support. The service providers meanwhile cast their tales around a *supernatural helper*, giving help to others to support a *voyage and return* journey rather than their own safe passage. They are bears with no habitat—no recognized natural place of being. The final

Once Upon a Time in Facilities Management. Patricia McCarroll and John Hassard, Oxford University Press.
 DOI: 10.1093/9780191943492.003.0008

group, in-house facilities managers, share the overarching typology of *supernatural helpers*, but under the category of *rags to riches*, where they identify with needing to receive supernatural help. However, this help is not provided by an external source, but instead represents a manifestation of their own resilience and hard work. They recognize the need for a fairy godmother within their profession but fulfil this role themselves.

The uniting concept across these tales, therefore, is a lack of recognition, including for FM's organizational legitimacy, with none of the tales enhancing a grouping's position from an external viewpoint. There is a consistent divide between the perceived function of FM as held by the industry and those who commission and use the function as patrons. Whereas the consultants and service providers interpret this as a source of alienation, in-house facilities managers take recourse to a narrative suggesting empowerment. This is a noticeable difference within the discipline and, as such, is a key finding from the research.

Turning first to the *tragic* story lines from the external FM consultants and service providers, there is clear evidence of anger and frustration within their story lines. The consultant stories revealed a distancing from FM, as they did not cast themselves directly into a character, indicating that the role of an FM consultant is one that has no power of voice over their own plot. This passivity is also evident within the tales of the service providers, albeit to a lesser extent. This supports the idea that 'we tell a story about what has been happening around us, which at least justifies and possibly heroizes us; alternatively, if we are feeling bad about ourselves, we many take the opportunity to tell a story which demonises, marginalises or even ridicules us' (Bruner 1990, cited in Sims 2003: 1199).

Both FM divisions are spectators in their own stories, with little control over their personal scripts. Their united motifs are those of impossible tasks and poor recognition/rewards, with this situation being exacerbated by limited power. As such, their stories project a reality that is not so much aligned to a strategic or professional approach, but instead reflect essentially disparate and disenchanted roles being played. With the understanding that organizations are driven by 'finding a place for everything and keeping everything in its place through imposing limits and maintaining boundaries' (Thanem 2011: 82), FM is arguably the embodiment of disorganization—a function whose presence indicates the chaotic nature of organizational life. In this manner, FM tends figuratively toward the illegitimate.

This is supported by the archetypal evidence, which not only makes the images and roles shared within the storied space of FM explicit, but also provides the wider organizational context to which FM takes recourse in its overall function as a service provider. Indeed, fragmented and blurry boundaries are the hallmarks of many contemporary organizations (Lewis and Kelemen 2002, Hassard and Morris 2020), crystallizing the need to understand the identity of secondary services, as presented in this research. Change appears constant, with the unquenchable demand for flexibility potentially causing organizations to lose their history and identity in the face of a new 'domain of an eternal, changing present' (Kostera 2012: 6). Within this environment, FM is associated with 'dirty' work (Cassell and Bishop 2014) in its link to janitorial operations (cleaning, gardening, general repairs, etc.) and the types of 'hard' FM (plumbing, heating, ventilation maintenance, etc.) illustrated in Chapter 2.

Indeed, within the consultant and service provider FM divisions, self-association with the *shadow* archetype reinforces this allusion to contamination. Representing an entity that is external to the contracting organization, the lived-experience tales within this research reflect their role as delivering 'change'—as outsiders who enter and upset organizational norms, thereby inducing a liminal state of uncertainty. In this manner, FM becomes the embodiment of the contracting organization's *shadow*, representing symbolically a 'process' perspective—one reflecting everything the contracting organization wishes to 'deny about themselves, due to the threat posed to self-image and self-understanding and, more generally, the need to be viewed in a favourable light by others' (Bowles 1991: 378). Consultant and service provider FM are fully aware of this, casting themselves into characters that are weak and forbidding, reinforcing not only a physical stigma of dirty work, but also one of social and moral stigma (Fraher 2017, Hamilton and McCabe 2016). This positions FM as representing the very element—dirty work—organizations try to subvert and hide, resulting in a lack of recognition and reinforcing FM's symbolic illegitimacy.

This lack of recognition is evident in the consultant and service provider stories, which display a 'paralysis of resistance' (Gabriel 2012: 1139). This is a phenomenon connected to what Gabriel has described as 'organizational miasma', a contagious state of pollution in which 'people become dispensable pawns, resources to be used, exploited and discarded' (2012: 1143). As noted, in the consultant and service provider tales this acts as the contracting organization's *shadow*—a situation of being aware of their negative role yet powerless to do anything but comply with wider organizational demands. However, the consultants' tales also reveal their own distancing from this metaphorical imagery of FM; allowing for detachment from the potentially contagious miasma that reinforces FM's sense of illegitimacy.

In-house FM shares with external FM emotions of frustration, but whereas the latter show anger in this regard, in-house FM display far more acceptance. Here, the paralysis of resistance is replaced by resilience over resistance. The lack of recognition that represents a source of illegitimacy still results in a liminal separation from the organization, but also forms a schism with the incumbent FM function; with the latter being portrayed as out of date and fading in the face of a new evolution of the discipline. This manifests into an internal focus and represents the main source of recognition—one which 'offers opportunities of liberation and empowerment' (Driver 2014: 92; see also Driver 2009) with the lack of organizational recognition being usurped in favour of internal strength. This is represented by their archetype representing not an external *shadow* but rather an internal *sage*, providing help to industry colleagues as well as wider service users and clients.

This empowerment allows in-house FM to complete their *rags to riches* journey and circumvent the tragic storied spaces of their external counterparts. The same enemy is shared across all the stories, but for in-house FM their recognition, and therefore enchantment, is not so much dependent on external recognition but internal respect held within the FM discipline itself, which is recognized as evolving for the better.

8.2 Key Finding Two: Benefiting from Betwixt and Between

With the nature of identity being recognized as fluid (Scott and Lane 2000), it is arguably more salient within a secondary service such as FM that operates within, and at the mercy of, wider organizational boundaries (McCarroll 2017). Therefore, the concept of liminality, and the metamorphosis this state offers, is a key finding, confirming the theory as presented in Chapter 2, and in support of the key finding of 'lack' as analysed above.

As highlighted, consultants and service providers assume the FM name and enter different organizations that are governed by their own internal norms and values. This aligns with the separation stage of liminality (as illustrated in Figure 2.4) across two boundaries. The receiving division of the contracting organization is separated from the remainder of the corporate body by the arrival of consultant or service provider FM teams, who represent the symbolic *shadow* archetype. They represent change, which is articulated vividly within the participants' fairytales; whose storied spaces reveal a sustained battle with a common enemy—of fear, unseen dangers, and transformation—which remains unconquered and largely intangible.

Simultaneously, as a role, consultant or service provider FM is also creating a liminal space—entering the receiving organization and thus an environment to which it must adapt or remain outside. This liminal state is exacerbated by the contractual nature of consultant and service provider FM, with their stories revealing an organizational state characterized by ambiguity, limited resources, and low status, all amidst client demands for immediate productive results. In this manner, FM consultant and service provider teams can find themselves *othered*: 'the process of casting a group, or individual, or an object into the role of "other" and establishing one's own identity through opposition to and, frequently vilification of this Other' (Gabriel 2008: 213).

By applying the concept of liminality, therefore, the different storied spaces occupied concurrently by such FM providers become discernible. The association with the *shadow* archetype carries strong elements of disenchantment and disconnection, as they potentially distance themselves psychologically from both the FM discipline and the main organizations they are contracted to, projecting a *tragic* storyline. However, the consultants' tales do not reveal a fully dystopian fairytale, despite the tragic story arcs: the liminal state allows for experimentation and reflection, a space to try on new and required identities.

In this manner, FM can act not only as a 'necessary evil' (Price and Akhlaghi 1999)—as proposed in Chapter 2 and supported by the re-imagined fairytales—but also in a manner to avoid the 'incapacitating ethos of self-criticism' (Gabriel 2012: 1140) that being 'betwixt and between' (Johnsen and Sørensen 2015: 321) can transmit. By acting as the *shadow* archetype, a negative, yet necessary, function is accomplished, allowing FM to facilitate the *voyage and return* journey of the client, as illustrated in the tales of *Goldilocks*. This is the symbolic initiation rite—completed by the client—which FM as a service provider is there to facilitate. But with FM having received no recognition, it is therefore not accepted, leaving the rite of liminality incomplete.

However, this unfinished state is counter-balanced by the acceptance of the liminal domain as a permanent condition—where continued growth and change can happen freely—one signifying great capacity and ability to metamorphose. For the FM consultants, this allows a state that creates distance from an FM designation and industry perception that is portrayed as a hindrance, thus preventing full recognition. It allows service providers to navigate the unceasing and fluid service demands made by an anonymous client—as 'versions of self are tried out and as a new or modified identity is sought' (Beech 2011: 289) in line with fresh contracts and contract amendments. This aligns to the role of magic contained in the fairytale genre, represented by an ability to suspend belief and be innovative, endorsing the theoretical position represented in Figure 8.1. Therefore, the unstable condition of liminality becomes not only a constant feature, arguably increasingly prevalent within a service provision industry, but one that is also necessary for survival. In the research, this is illustrated in instances where the fairy stories of consultant and service provider FM do not fully lead to a conclusion, but rather remain '*to be continued*'.

Correspondingly, the in-house FM function also reflected elements of a liminal state, although—in contrast to the other FM classifications—this role category is naturally not external to the parental organization. This is, however, purely a contractual phenomenon and elements of liminality are actually more prominent within the in-house FM tales. Not only do they disclose a notional separation from the parent organization, but there is also evidence of an abstract departure from the wider discipline of FM. Where the consultants find this enabling—facilitating an unending liminality—for in-house FM the stable state of aggregation is not only viewed positively but also one to be achieved through a process of internal-focused recognition. As such, in-house FM represents the only re-imagining of a fairytale that is, in this sense, completed, indicative of the utopian association of the genre.

Taken from a collective view, therefore, the concept of liminality illustrates a quality of metamorphosis that is arguably imperative for survival within modern corporate affairs. This reflects a situation where managing boundaries is 'not necessarily in the traditional sense of controlling them—but rather of imagining, understanding, sharing and influencing them' (Huemer et al. 2004: 63). Within a secondary service such as FM, this position is not only emphasized in a lack of legitimacy, but is also arguably compensated by it, as demonstrated in key finding one. As such, this metamorphic capacity—revealed by liminal states and carried symbolically by the re-imagining of fairytales—suggests for FM professionals 'an ability and a willingness not merely to adapt to change but to redefine themselves radically, to metamorphose into new entities' (Gabriel 2005: 13).

It is established thus that the profession as a whole is in a liminal state, with no tale demonstrating an acceptance into the main organization that FM, as a secondary service, is aligned to. The re-imagined fairytales illustrate the different positions within the liminar state (Figures 3.1 and 5.1). Consultants recognize the permanency of the liminal state and their place within it—not only accepting it but using it to their benefit. This is shared by the service providers, but to a lesser extent as the detrimental impact of organizational illegitimacy is greater; as they fight to 'maintain an identity stable enough to warrant sustained involvement from others such as suppliers, customers, regulators . . . while remaining alive to a shifting multiplicity of standards and expectations' (Zundel et al. 2016: 218). In-house FM professionals, however, are most affected, with organizational illegitimacy being acutely apparent through their lack of a collective acceptance by 'members that their organization [discipline] is, to some extent, desirable, proper or appropriate' (Brown and Toyoki 2013: 875).

Returning to our analytical origins, it can be argued that the works of the Grimm Brothers highlight a key sociological question raised by fairytales: 'how can one learn what must one do to use one's powers rightly in order to be accepted in society or recreate society in keeping with the norms of the *status quo*?' (Zipes 1983: 57, italics in original). In this analysis, we have noted how concepts of liminality and metamorphosis are cornerstones of stories and tales of professional sense-making. What is more, we have described how such concepts bring to light the symbolism of organizational meaning as recognized within this example of a secondary service. Such reasoning offers new insights into the main FM delivery mechanisms and the nature of their professional empowerment—offering evidence of the functional benefits of liminality despite the fact that this may function at a wider cost in terms of disciplinary legitimacy: 'Our organizations are not interested in listening to the Other. They are interested in labelling, classifying, and managing them, in short, incorporating them into their own narratives' (Gabriel 2004: 630). With new understanding of the impact and usefulness of liminality, related concepts of metamorphosis, legitimacy, and enchantment can be highlighted, as illustrated in Figure 8.1. By emphasizing the presence of what are considered liminal spaces, and linking their associated social, organizational, and political traits to the FM profession, the precarious, anonymous, and ambiguous nature (Lucas 2014) of FM is uncovered, opening the way for an understanding of why FM—without a stable identity—has persisted as an industry for nearly half a century.

8.3 Key Finding Three: The Family Business of FM

As discussed in Chapter 5, narrative analysis of the three FM delivery mechanisms—consultant, service provider, and in-house—has resulted in different story arcs and archetypes being manifested. This is opposed to revealing a 'grand plot' typology that would 'establish rules for the construction of appropriate, or admirable, plotlines' (Kociatkiewicz and Kostera 2016: 325). This line of analysis would indicate, therefore, that FM is not so much a united discipline as one comprising various divisions, which consider

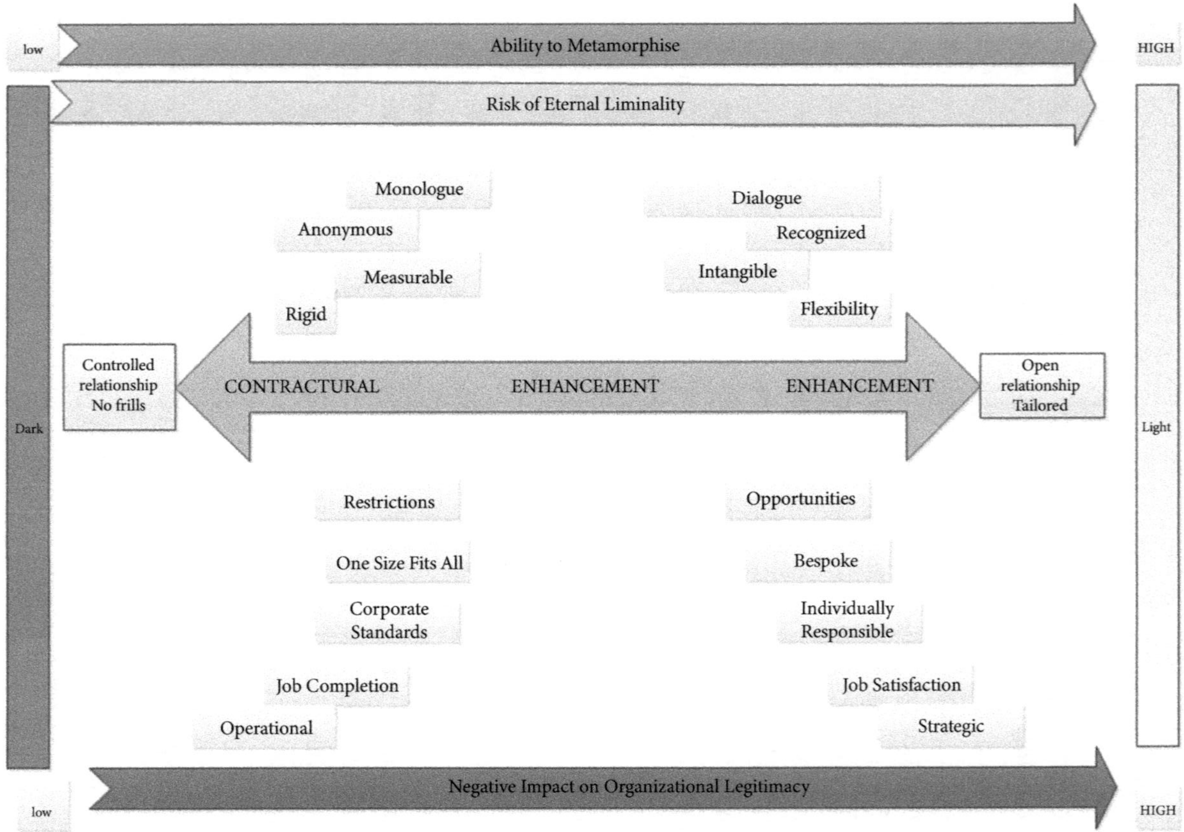

Figure 8.1 The enchantment of FM
Source: Authors

themselves as essentially separate organizational entities, and hold different opinions on the functional roles of one another. However, this finding also suggests they tend to present an accepted and collective 'internal' view of their operations and thus are united to an extent through co-dependency. Such internally recognized legitimacy manifests itself to support an integrated storied space of the 'family', with this arguably bolstering the six attributes of professionalism outlined in Table 2.1.

As seen in Chapter 6, all three chosen fairytales (*Hansel and Gretel*, *Goldilocks and the Three Bears*, and *Cinderella*) share the characteristic of kinship. In the consultant's re-imagining of *Hansel and Gretel*, for example, the family theme was specified by C3 as the reason why this fairytale was originally selected. In this tale, FM was the family, bonded together under a shared title; although here in a scenario which ultimately had become dysfunctional, with little or no leadership being shown; instead, end-users were cast as Hansel and Gretel, who need further guidance and support from consultant FM. Meanwhile, service provider FM took recourse to the symbolizing of the three bears as a nuclear family, here illustrating the hierarchical nature of central roles; with father, mother, and baby representing different levels of FM experience, as opposed to making reference to acquired professional rank. While finally in-house *Cinderella* illustrated both the fairy godmother bringing service provider FM into the organization's family and the 'princely' role of both service provider and consultant FM in the tale of in-house FM.

The reoccurring family motif, therefore, is present in all stories when FM delivery mechanisms empirically discussed the roles and actions of others. This finding illustrates the overlapping and occasionally contradictory power relationships beheld metaphorically within an overarching family dynamic. This represents a symbol of unity and stability, but one perceived tacitly within, and qualified by, the recognized bonds of an allegorical household relationship. Thus, the notion of the family unit suggests a pre-existing and unspoken structure—one with appropriate supplementary roles and accompanying expectations.

In respect of research aim two, however, the parts that the different delivery mechanisms of FM cast both themselves and their FM counterparts into reveal no overt heroes. This appears contradictory to the superhuman rhetoric that has arguably continued to underpin the storied spaces of organization studies for many decades, and notably as reflected in the grand narrative of perhaps the best-selling book on management style and corporate control of all time—Peters and Waterman's (1982) *In Search of Excellence* (see Höpfl 1995 and Collins 2013 for critical commentaries). For the current investigation it is interesting to note that fairytale heroes tend to serve as role models of courage and leadership—a source of admiration (Forster 1999), which is translated organizationally to 'personify the values and epitomise the strength of the organization' (Jones 1996: 58). The artificial folklore of FM, however, would suggest that FM is only a hero in that it holds notions of followership, and therefore fellowship, as central to its family ethos; that is, underlying philosophies whose correlated explicit actions are equally devoted to helping or even saving one another within FM's functional roles.

By highlighting the motif of the 'FM family', this analysis also serves to illustrate the nature of internal and external demarcations in the profession. Here FM is the core family, with associated and recognized 'shared values' presenting a clear 'understanding why they are alike and together, distinct from others within their field, and yet close enough to define themselves in relation to it' (McCarroll 2017: 293). Above all, it is the nature and form of client roles that helps to define FM—insofar as the discipline, as a service provider, is dependent upon them.

Nevertheless, FM can metaphorically cast client roles as an external entity—one that throws the stable family unit into a state of liminality. In so doing, the notion of the client role can be portrayed as dominantly negative and cast in the villain role. As the FM hero does not exist in the wider organizational story, it can never save the client. In turn, the client casting process also illustrates that clients do not expect to be saved by FM actors, who have cast themselves, for example, as weak (Hansel and Gretel's father), misunderstood (Goldilocks's bears), or even as self-rescuing (in the case of Cinderella). This supports the second key finding, in that FM remains positioned in an eternal, liminal, space—part of a client's tale, but one facilities managers are not empowered to finish.

There is also here a suggestion or perhaps taint of 'dirty work' (see Simpson et al. 2012), which conceptually is arguably the obverse of the hero perception. An association with dirty work can be the source of wider organizational exclusion (Cassell and Bishop 2014), but arguably also acts as a reference point to enhance occupational identity and cohesion (Fraher 2017). In this context, the FM family creates a dynamic of an occupational group that facilitates disciplinal divisions such as consultant, service provider, and in-house—divisions that are malleable yet recognized within the family-based unit; divisions representing who organizationally is 'good' and who is 'bad' and reflecting how particular divisions need to act within a specific workplace.

In this way, there is a sense of belonging to a wider group—a shared sense of kinship and communal affiliation to the FM appellation. There is no evidence of internal ambiguity concerning the nature of FM and the different organizational elements that comprise it. For example, FM consultants applying for contracts are fully aware of their professional identity. However, they display characteristics of a liminal organizational state—not because of professional uncertainty on their part, but due to the uncertainty displayed by the contracting body and its understanding of FM. 'There was a lot of mistrust. People felt that nobody (in the organization) knew what they were supposed to be doing. There was no leadership' (C4).

Therefore, the family motif allows a shared ethos to be illustrated, even if the stories are different. This presents an ideology somewhat indicative of a 'community of practice' (CoP) approach (see Lave and Wenger 1991). These are communities that share history, creating similar outlooks but 'their members do not have a shared task or agenda; they are not accountable to anyone but themselves' (Gabriel 2015: 98). This internal view of accountability is therefore arguably linked to the wider issue of organizational illegitimacy of key finding one. Internally, facilities managers recognize their actions as legitimate insofar as they are 'desired, appropriate and acceptable' (Landau et al. 2014: 2), but this rationale is not necessarily always recognized by the client, as emphasized frequently in the role of the latter as villain.

In conclusion, although the professional folk of FM occupy diverse storied spaces, all nevertheless 'share a common orientation and overarching purpose, face similar problems, and have comparable experiences' (Meyerson 1991: 133). Given this logic, appreciating and understanding the stories of the three FM delivery mechanisms helps us to see how they fit together in terms of larger organizational narratives. This argument indeed forms the basis for developing the form of narrative analysis developed and demonstrated in the *artificial folklore* method of Section 8.4.

8.4 Key Finding Four: The Fairytales of FM

In developing an artificial folklore approach to narrative analysis the aim was not to create an ontologically and epistemologically 'real' story but to generate ideographically substantial, convincing, and authentic 'ghostwritten' tales of FM. In essence, accessing the diverse approaches that ultimately comprised the artificial framework approach allowed a 'way of making the familiar strange, and making the strange familiar' (Knowles and Cole 2008: 11)—a method offering distinctive cultural insights into the multifarious storied spaces of FM. The three tales presented within this section provide a methodological lens 'to see that which is agreed upon by all organizational members, that which is shared only within certain groups, and that which is fragmented and ambiguous' (Søderberg 2003: 5). In developing this method, we have accepted that: 'Nobody can write a new fairytale; you can only mix up and dress up the old, old stories' (Lang 1910 as quoted in Ashliman 2004: 158). As such, each of the fairytales generated is open to interpretation. The philosophy suggests recognition of overlap in certain areas of analysis, notionally to ensure that the artificial framework presented allows for flexibility both in terms of folklore method and views expressed by participants.

The first tale presented is that of the consultant's storied space, which is a re-imagining of Hansel and Gretel under the *supernatural opponents/overcoming the monster* category. As FM was allocated the role of the father, the tale was written from that viewpoint. This altered the original tale significantly, moving the character of the father from a shadow figure, present only at the start and end of the tale, to that of the main protagonist. The resulting tale, therefore, only alludes to the importance and experiences of Hansel and Gretel, reflecting the distance from the discipline that this division of FM purposely creates. Meanwhile, the stepmother role symbolizes both the unhappy relationship between FM and the contracting organization, and the fluidity of this role association, with its unstable power dynamic. Despite being central to the story categorization, the *supernatural opponent* that is defeated in the tale remains essentially an afterthought—one that is not to be acknowledged, notwithstanding the threat. In direct reference to the re-imagining, the consultant's tale is eternal; it remains stuck at a metaphorical crossroads, thus employing the ending of Grimm Brothers' infamous *Golden Key* tale (see Zipes 2014a).

The second tale presented is the re-imagining of *Goldilocks* from service provider FM. In contrast to their consultant colleagues, this tale centres on the story arc of the *supernatural helper*, but also referencing a *voyage and return* sub-category. This is ultimately a tale of liminality, where the supernatural helpers are principally in the guise of the FM bears, the givers rather than receivers of supernatural help. However, this help is not wanted, or indeed recognized; although it is essential for the protagonist, in this case for Goldilocks's immediate survival and long-term growth. The overarching archetype of *sage* for the service providers suggests their role in guiding the clients. Despite this being perceived as a negative experience by the clients, the role of the FM *sage* allows the client liminality to be resolved. This is at the expense of service provider FM, which must remain in the liminal space of the woods—as its role is to help the client, Goldilocks, to learn and return home; so there they must stay to be effective, regardless of the adverse association.

The final story presented is the *Cinderella* of in-house FM. This is also within the category of *supernatural helpers* but aligns to the more positive *rags to riches* typology. The re-imagined tales of in-house FM revealed storied spaces that ran across a broad narrative spectrum, showing the wide variety of functions within FM. Here the storyline of change and mergers is dominant—reflecting a limited variety of characters in the casting but emphasizing the liminal state, where new identities are created, recreated, and experimented with. This quality of metamorphosis influences the fluctuating boundaries of FM, arguably affecting the discipline's ability to achieve a recognized and therefore legitimate and stable state, suggesting considerable flexibility in terms of growth and evolution. This is clearly evident in the tale from IH4, not only in this re-imagining of being the supernatural helper to a service provider security guard, but also in reflecting on the career that FM provided him—the ability to start at the hearth and rise to run the palace by being offered a possible directorship. Only one tale within this category did not carry a positive outlook. This re-imagining (IH8) left Cinderella by the hearth in perpetuity, awaiting the time where her skills are recognized. In terms of FM, it is noteworthy that she does not change profession, but accepts her fate and settles down for the wait, in the knowledge that this time will come, supporting the liminal analysis of key finding two.

Taken together, therefore this collection of fairytales serves to 'depict and interpret vitally important yet not ostensibly visible organizational aspects and processes' (Kociatkiewicz and Kostera 2014: 3) across the three provision areas of FM, completing the final stage of the *artificial folklore* framework.

8.4.1 The FM Consultant's Fairytale

Once upon a time there was a poor father who lived in a tiny cottage at the end of a forest with his children Hansel and Gretel. It was a time of hardship across the land, and although he was skilled, there were not enough resources to go around and his small family were now starving.

To try to improve his lot, he remarried but he quickly became overwhelmed. His new wife was rich but did not like the existing arrangements with his two dependents, his children, and pressured him to break this long-standing agreement for the good of them all.

'How can we feed the children? We have enough for two not four,' the woodcutter cried. His wife, who managed the household now, thought he was a fool and demanded that the two children be left in the woods to make their own way in the world. This made the woodcutter very unhappy. He had a duty of care and wanted to honour the commitment he felt he had made, but his partner did not understand this. However, as she now controlled all the resources, he reluctantly carried out the orders, and tricked his two children to follow him into the woods where they quickly became lost without him.

Broken-hearted, the father reconciled his decision. 'At least I have food on the table and clothes on my back now,' he thought. He spent his days completing the basic household chores that his wife ordered him to do and tried not to think about it too much, as no good would come of that.

Then one day, Hansel and Gretel found their way home. The father was overjoyed to see them, as all the chores were increasingly hard to complete by himself. However, his wife was horrified that instead of making their own fortunes, the children had followed the path home once again. She did not care for this relationship and viewed any long-term commitment in this household as foolish.

And so the poor woodcutter took his children still further into the woods. He thought they knew what he was doing but they did not speak about it. 'There are terrible things in the woods,' he thought. 'But the children need to learn that things always change, and they need to adapt. It is for the best.'

As the days passed, the woodcutter grew tired of all the demands his partner made upon him, and resources were once again scarce. He wondered if he had acted too rashly as, although getting rid of the children helped in the short term, he missed them terribly and he was all alone when his new wife abruptly left. Knowing he could not survive by himself, the father decided to seek a new alliance. He had heard of a marriage fair in the next village and travelled there with great hope.

There he met someone who reminded him very much of his previous partnership. Although this had failed and he had suffered the loss of his most beloved children, the familiarity appealed to him. However, there was also the offer of a merger with someone who was completely different; indeed, he had never come across anything like this before. 'One will solve all my problems today,' he thought. 'The other will take time and is riskier, but if it worked, I could look for my children and we would live happily ever after.' And so he sat down to think about this, and so now we must wait until he makes up his mind. That's when we'll see what his future holds.

8.4.2 The FM Service Provider's Fairytale

Once upon a time, there lived a family of three bears in a cottage in the woods. They had chosen their home carefully, slightly off the path and hidden under the trees. Father Bear, Mother Bear, and Baby Bear were very proud of their house and their efforts to maintain it, and they ran a B&B from it as their main source of income. Talking bears were still unusual and so they had to try extra hard to secure reliable bookings. Father Bear also worried about Baby Bear. He was very innocent and once had come home with all his belongings gone and so Father Bear was doing his best to educate his young protégé on how things were done and will always be done.

Every morning, Mother Bear made the family some porridge. And with every morning, it was too hot to eat, and so the family took a walk outside to let it cool. Soon a lost little girl came across their house. She didn't know she was lost, and so she was unafraid and when she saw the door to the Bear's house was open (as it always was), she went straight in.

There she saw three bowls of porridge. She wasn't hungry until she saw them, but her tummy started rumbling and so she took a big scoop up from the first, biggest bowl. She howled and stamped her feet, as the porridge was too hot and burned her mouth. She grabbed the second medium-sized bowl. This one she spat out across the table—it was too lumpy! The last, and smallest bowl was just right for her, and so she ate it all up.

Goldilocks then saw three chairs. She sat in a very big chair, but it was too high! She sat in the medium-sized chair—it was too hard! Then smallest chair was just right but she squealed as it collapsed under her weight. She was too heavy for it and it broke into many pieces.

Next Goldilocks went into the bedroom. She had looked all around the house, and she was tired. She climbed up onto a very big bed. It was too hard! The she climbed into the medium-sized bed. It was too soft! Finally, she crawled into the smallest bed, which was just right, and she fell fast asleep.

Soon the three bears came home for breakfast. 'Who has been eating my porridge?' they all asked. 'Who has been sitting on and breaking our chairs?' Father Bear looked at his very big bed. 'Who has been lying on my bed? He asked in a very loud voice. He tried to think if they had any bookings that day. He did not think so. Whoever it was must be tired from a long journey and needed to eat and rest. Just as he was thinking about this, Baby Bear found their guest asleep on his bed. As he went to wake her and explain about the guest accommodation, she jumped up in fright. One by one, she looked at the three bears. They were big, scary, and growling at her. Now fully rested and fed, she rushed to the window and ran quickly away into the wood.

Father Bear watched her go. Although they now knew what she liked, what chair and bed suited her best, he knew she was yet another customer who would not pay her bill, never mind the damage she had caused. He was a little sad that he would never see her again, but also confused. She had been the fourth customer that month that had acted this way.

8.4.3 The FM In-house Provider's Fairytale

In a time when wishing still did some good, Cinderella sat by her hearth. She was dirty from the ashes, but the fire kept her warm. She had built it herself and so was content it would last hours, providing her with warmth and safety. Times were hard, as they always were, but her stepsisters were excited as the prince had come of age and was looking for a bride, bringing with it promises of a new age of prosperity.

Cinderella did not know her stepsisters well. She wanted to, but they stayed in the other, richer area of the house and only came to see her when there was something that needed fixing or ordering. Without them, Cinderella would have no place by the fire at all, but they were so bad tempered that they looked ugly.

Then one day, her stepsisters burst into the kitchen when Cinderella was sewing, as a grand ball had been announced. They needed her to make their dresses, and practice doing their hair and makeup so it was just right to catch the prince's eye. All the beautiful young girls in the country had been invited but Cinderella was not. As she was only seen in rags, working in the kitchen, everyone thought she was her sisters' maid. There was no question of Cinderella going to the ball, but she often dreamt about it in those weeks, when the sisters would tell her off for being lazy. 'You need to see **me** *now', they all said at once.*

As they left for the ball, Cinderella sat down at her place by the hearth and cried as if her heart would break. Suddenly, a kind voice startled her. 'What's the matter, my dear?' There stood her fairy godmother, smiling. 'I would like to have a beautiful dress and go to the ball and dance with the Prince,' Cinderella said. 'And so you shall,' said the fairy godmother. 'Dry your eyes and do exactly as I tell you.'

And so Cinderella ran to the garden and picked the biggest pumpkin she could find, but try as she might, the fairy godmother could not get it to change into a coach. The mice refused to change into horses, and the coachmen remained rats. She tapped Cinderella with her wand, but her clothes remained old and dirty. 'I am sorry my dear,' the fairy godmother said. 'But I really do have to go now, I have lots of other people waiting.' And with that, she disappeared.

With future promises of change ringing in her ears, Cinderella gathered herself up and went out to find the local seamstress, who had taught her to sew. Glad to see her old friend, the seamstress gave her the best dress she had. Next, she went to the mechanic, who Cinderella had sent her sisters to when they wanted a lift. He smiled when he saw her and ordered his best car as she was one of his best clients. Lastly, she went home, changed, and got herself ready with all the skills she had learned over the years.

When Cinderella arrived at the palace, she looked so beautiful that the ugly sisters did not know her. They thought she must be a princess from another country. They never thought of Cinderella, for they believed that she was sitting at home, by the cinders.

The prince thought that he had never seen such a beautiful princess. He danced with her all night, to the point where her shoe broke. He was amazed as she calmly sat down and began to fix it, and as she did so, they talked and talked about the land, the castles, and the people. The prince recognized her talent, and what she could do if enabled with his resources and reputation. And so together, they formed their own partnership and lived happily ever after.

8.5 Coda

This chapter has identified the key findings of this research, culminating in the *artificial folklore* framework by presenting ghostwritten fairytales for each of the three delivery mechanisms of FM. In this analysis, the postulated notion of lack is seen to unite the discipline in terms of manifesting the necessary but vilified *shadow* archetype. Similarly, the notion of liminality is evident throughout analysis of the three provision types; although being applied and positioned in different measures, with this leading to a range of diverse yet accepted roles within a family structure that functions despite the difficulties presented by a client villain. The tales relate to a shared understanding of the essential nature of FM—that it is a hidden, unrecognized, and largely invisible discipline; one bound together professionally mainly by a motif of mutual understanding.

This highlights the roles and therefore identities that FM professionals cast themselves into—the storied spaces they consider themselves part of and which create a set of reciprocal expectations; the plots and associated transformations that are at the core of their 'stories' (Robichaud 2003). It also represents disconnection between the images that FM as a discipline projects and the identity of those who carry out the tasks of the function. In this manner, the ghostwritten fairytales presented above are both a reflection and a product of contemporary FM as an example of the wider service provider market.

Finally, the methodology developed in this study has hopefully gone some way to allowing the rich cultural and organizational complexity of FM to be revealed, with each tale illustrating a different discipline plot line and set of relations within a storied space—the latter being analytical domains that may have been obscured by more conventional research formats and approaches. The storied spaces analysed reflect the diverse qualities of participants' experience, the complex nature of FM as an organizational function, and the diverse political struggles in which the discipline is embroiled. The approach of *artificial folklore* therefore has facilitated insight into how FM is experienced subjectively and culturally in everyday practice, as opposed to how it may be relayed ideally in reified form in strategic plans or corporate communications.

9
Conclusion
Happily Ever After?

This research has analysed the secondary service of FM as an example of a contemporary non-core service utilizing an *artificial folklore* methodological approach. It has taken the stance that fairytales 'select that which has become relevant in a community to inform members of that community what has become crucial for adaptation to the environment in the most effective manner possible that might be entertaining and instructive' (Zipes 2014b: 13). The multi-disciplinary approach expanded the idea that fairytales have been revised, retold, re-imagined, subverted, and invented as they respond to social values and needs over time, and focally as applied to a contemporary organizational setting. At this point, the approach developed must now be reviewed in terms of its theoretical and practical value. As such, this chapter considers the contribution to both knowledge and the discipline of FM. These issues are assessed alongside the limitations and constraints that were uncovered during the investigation. Finally, further recommendations for research are presented and conclusions reached.

9.1 Contribution to Knowledge

The contribution of this study firstly lies within the creation of the *artificial folklore* framework. This represents a multi-disciplinary study influenced by an arts-based approach (Kelemen et al. 2018) that intertwines folklore, literature, and narrative theory with organizational theory. By doing so, the approach reveals the organizational status of this secondary service and the character of the evolving divisions within it. The method has served to capture not only perceptive and insightful stories and archetypes illuminating the nature of the FM discipline but, in so doing, also to provide a deeper understanding of the organizational life-worlds of its practitioners: 'It is argued that by paying attention to the symbols, tales, legends, and myths that organizational members use to describe their experience, the researcher can

Once Upon a Time in Facilities Management. Patricia McCarroll and John Hassard, Oxford University Press.
 DOI: 10.1093/9780191943492.003.0009

tune into operative dynamics that would otherwise remain very covert and inaccessible' (Smith and Simmons 1983: 377).

Although the use of stories as a means of explaining the nature of organizational life is a generally recognized investigative approach (Boje 1991a, 1991b, Boje and Baskin 2011), folklore as a specific research tool is less apparent, despite the overlap with many aspects of Organizational Behaviour (OB) (Price and Whiteley 2014). As fictional accounts created by researchers remain 'highly unusual in management studies' (Whiteman and Philips 2008: 289), this research aims to broaden the base of organizational studies by exploring alternative textual forms that 'function to sustain the discourse of organizing' (Philips and Zyglidopoulos 1999: 594). Simultaneously, it also aims to provide unconventional approaches that can test 'our attachment to entrenched academic traditions surrounding the production, representation and consumption of research' (Whiteman and Philips 2008: 297), by introducing folkloric principles.

Folklore in this study is positioned as 'a system, orientated toward conventions sanctioned by a cultural community' (Tatar 1987: 66). As such, its use within the present approach to social research provides a novel means to capture and understand the dynamics of increasingly complex and fluid organizational boundaries (Rippin 2013, Kociatkiewicz and Kostera 2014). The methodology developed here helps answer the call from Czarniawska and Mazza (2003) for contemporary organizational research to become as flexible and emergent as the workplaces it studies. With progressively blurry occupational boundaries, fairytales offer the capacity to capture the essence of mutable and capricious workplace phenomena, for 'the fairytale is in a perpetual state of becoming and alteration. To keep to one version or one translation alone is to put a robin redbreast in a cage' (Pullman 2012: xix).

We have also argued that the use of fairytales within organizational research is innovative and exceptional. This study aims to recognize the potential of fairytales as a medium of social explanation—a prospect which raises 'the question of individual autonomy versus state domination, creativity verses representation . . . [for] . . . just the raising of this question is enough to stimulate critical and free thinking' (Zipes 1979: 21). Where folklore can expose the principal conventions established within an organizational community, the specific medium of fairytales serves to capture their emergence and evolution at more nuanced and projective levels, for they draw on 'the double movement of language between literal meaning and figurative expression to fashion stories that dramatize psychological realities' (Tatar 1987: 82). The clear narrative structures and palpable fictional characters provided by fairytales sanction a recognized conduit to the ordered constructions of storytelling, allowing events to be exaggerated in order to build a tangible evaluation of their socio-cultural reality (Dégh 1972). Utilizing Propp's (1968) framework, for example, Morrell and Tuck (2014): in a study of 'governance, tax and folktales') argue the tales we absorb at an early age create conceptual catalogues we instinctively deploy to interpret more complex terrains in later life. In wider fields, such as anthropology, psychology, and sociology, writers such as Shokeid (1982), Bettelheim (1976), and Baruch (2009) have employed fairytales to construe the dynamics and complexities of social and relational life, utilizing them principally as an inductive sense-making tool.

In the present investigation—predicated on developing *artificial folklore*—the combining of factual information and fairytale narratives has positioned the telling of stories not only as a form of data, but also as a methodological practice. In this manner, fairytales are both the process and the product, being created by the methodology and then studied within the data analysis. This approach presents an organizational reality often missed, or which is inaccessible, through the use of conventional methods of data gathering and sources of information (Gabriel 1991), allowing 'access to the unconscious yet projective images of what the [discipline] meant to the managers' (Rhodes and Brown 2005b: 169).

The artificial folklore approach therefore has provided admission to a 'backstage perspective' (Van Mannen 1973: 420) of the FM function and discipline—a view of organizational practice that pierces the polished corporate veneer of untrammelled imageries of professionalism in secondary services. Recognizing and incorporating poetic licence has allowed for accessing a greater degree of honesty, depth, and reflection than may have been realized by relying merely on lived-experience tales and the use of standard research instruments. The artificial folklore approach allowed scope to capture 'issues such as absurdity, irrationality, uncertainty and disorder which are often overlooked in organizational research' (McCabe 2016: 25). By doing so, it highlights the actions and status of actors who reside in the 'netherworld'—those who both remain in and archetypically represent the organization's *shadow*:

> Our organizations themselves appear to create more and more borders and to patrol them with ever greater vigilance; these act as boundaries to movement, to communication, to expression, to narratives and, of course, to human contact. We are creating more and more Others, whose stories remain unlistened to, whose voices are unheard, whose Otherness is ever more untouchable. (Gabriel 2003: 620)

As developed here, the artificial framework provides an approach that can be considered generalizable or transferable across the organizational research landscape. Transferability here relates to the extent to which abstracted factors and tacit influences can be applied or added to our understanding of more material and measurable forms of organizational relationships and understanding how the former can be developed and strengthened as new theory is produced and matures, principally through further—potentially triangulation-based—research investigations (see Modell 2009). The specific steps that form the framework for such conceptual and methodological developments are detailed in Appendix 4.

The issue of participant engagement in this research can also be considered a contribution. With modern workplaces becoming increasingly time pressured (Hassard et al. 2012), completing organizational research can be problematic in terms of securing time with participants. Aligned to a wider call for creativity and innovation within industry, this research was therefore positioned and promoted as 'interesting' in that it potentially subverted audience assumptions (see Gabriel 2013). Although there was a risk that participants might find the folkloric approach confusing, or at worst, juvenile, the use of familiar stories and characters appeared to engage those who donated their time to the research. Indeed, this element of curiosity (Kelemen et al. 2018) not only secured a good number of volunteers, but many commented on the enjoyable nature of the research, which in turn provided a rich variety of data. One participant commented, for example: 'It's strange, when I first . . . got your e-mail and stuff, I thought you were nuts! . . . But there's method in your madness . . . I think it's great, and I would love to see the end product' (IH4), while others said: 'That was really interesting, gave me lots to think about' (C4) and 'I enjoyed that; I hope you get something out of it. I went down the bear route—I like that!' (SP1).

In carrying out the research we recognized that 'working with ethics involves realizing that we do not know how others will respond to and/or interpret our work' (Adams 2008: 179). With this in mind, we felt the analogical approach to data collection adopted helped ensure there were no major ethical issues of concern within the research process. Moreover, this principle of ensuring ethical research practice was supported tangibly and explicitly by use of a participant consent form and the assurance of full subject and organizational anonymity throughout the investigation.

Lastly, the impact of the main researcher in terms of providing input and contribution to the project is noted. As stated earlier in the sections on the research approach, the main researcher has a background in FM both as a practitioner and as an academic. It is acknowledged that this may have influenced interpretation (as addressed in the 'limitations' section, Section 9.3), but hopefully not in a manner which destabilizes the underpinning Jungian influence outlined in Chapter 3. Drawing on his analytical psychology approach, Jung practiced a dialectical procedure that emphasized the equality of all participants—metaphorically journeying with them together—as opposed to a logocentric doctor/patient relationship, with associated asymmetrical power relations and disproportionate agenda-setting arrangements. Instead, experience gained as a former practitioner facilitated general understanding of the FM function and professional roles and recognized a need for further research in this field, specifically focusing on capturing the voice of the FM managers themselves: 'Only the wounded physician heals. It is very important not to know all the answers. Often, we don't know . . . [I]t is greater to the patient when he discovers the answers himself' (Jung, as cited in Bennet 1982: 32). Arguably, this allows the contribution to knowledge to be richer, supporting a 'unique signature that spans the scholar–practitioner divide and produces novel results' (Fraher 2017: 139).

9.2 Contribution to Practice

The focus of this study has been the practice of secondary services—those that support the main organization and, as such, do not directly generate profit. Support services are thus the most vulnerable in terms of stability and permanence within an organization, notably due to an escalated outsourcing risk in contemporary economic climates. This fluidity and positioning as *non-core*, therefore, directly impacts the identity of these services, as illustrated in identity issues facing related outlying disciplines, such as HRM (Keenoy 1999), IT (Wilcox 1992), project management (Hodgson and Paton 2015), and hospitality management (de Bruijn et al. 2001). In this study, FM is positioned as representative of peripheral organizational practices (Westwood et al. 2014), arguably being one of the most salient due to its eminent, unadorned, and ongoing legitimacy issues (Drion et al. 2012).

Using the imagery of a fairytale to introduce the various professional facets of such a corporate sub-discipline is viewed as an innovative approach to organizational analysis. Thus, the approach has aimed to make a range of contributions to understanding such organizational support practices. The familiar imagery of the fairytale has allowed the exploration of what practitioners themselves project on to functional areas, including description of professional archetypes formed within each narrative account of undertakings within the discipline. The stories generated, therefore, not only allowed formal and tangible features of FM practice to be represented, but also for informal and intangible views of FM's management and organizational status to surface. Such an approach also permitted participants to assume the role of narrator to illustrate how they see themselves in their own storyline, thus exploring critically the conventional and habitual images the practice promotes of itself in creating and reproducing functional narratives (see Kociatkiewicz and Kostera 2014). This approach, we argue, could be applied in studying a range of other managerial and organizational disciplines, that is, to explore more fully the opportunities arising from this new form of research methodology.

The creative approach as utilized within this research also relied on the ability of the storyteller to connect to the brief and think slightly 'differently' (Hassard et al. 2008), using imagery and metaphor to create their own fairytale analogy. With the role of narrative fiction being connected to the 'development of imagination for action' (Czarniawska-Joerges 1995: 17), this approach contributes to the further development of creativity and innovation within practice. This is particularly salient in service sectors, where there is increased competition and therefore the need to gain and maintain a competitive market status (see Cui and Coenen, 2016): 'It is essential for the established supplier companies to prepare themselves for a future that brings with it immense competition, globalization, new product development and generation of an environment that insists on business innovations' (Goyal and Pitt 2007: 59).

Czarniawska (1998) suggests that using a narrative approach can facilitate organizational reinvigoration, allowing the recognition of the often fragmented and hidden knowledge to be supported with metaphorical application. A case in point is her work on the Swedish public administration, which utilized screenplays to illuminate paradoxes, pinpointing crisis points that allowed the process of organizational change to begin. Pässilä et al. (2013) echoed this argument in respect to forms of organizational innovation. With FM being chosen as an exemplar secondary service—due to its ambiguous identity—adopting more of a storytelling approach within the practice, may allow a refocus on relationships, on intangibility, and therefore reinvigoration, either within liminal organizational states (in the case of FM consultants) or by acceptance into the wider organizational landscape (in the case of in-house FM). By capturing stories of the practice, past, present, and future (Rhodes and Brown 2005b), the symbolic meaning of FM can be understood in line with changing organizational needs: for example, 'narratives can be seen as orientating inscriptions of past performances and as scripts that provide staging instructions for future performances' (Küpers et al. 2013: 84).

Earlier, the analysis of Chapter 2 pinpointed the risk of FM, as a secondary service, being judged by what can be measured. This shapes not only the external perception of FM as a cost centre, but also how FM, internally, forms itself to meet this expectation. By using a storytelling approach, it empowers practitioners to create, as opposed to follow, a wider organizational storyline:

> As we hear a story, the events are not meaningful for us until we have read ourselves into the story. This is how a story, and the telling of it, functions in a fundamentally different way for us than does analytic science. Analytical science gives us events that are objective fragments of reality and leaves us detached from them; the story always gives us events that are intended to be coherent and meaningful to us, something that cannot happen unless we become involved with them. (Hummel 1991: 36)

The contribution to practice in the research approach, therefore, functions in a way that captures the formation of FM's 'tale'; in so doing, compensating for the lack of any official organizational narrative (Boje 1995). The understanding that an employee's identity affects how they undertake their role (Butler et al. 2014), is thus applied at a wider level for the FM discipline. By capturing these tales, insight into the lifeworld of FM is facilitated, such as the perception branding highlighted by the casting process of artificial folklore: 'The individual in the organization who is assigned a fairy tale role might be branded by that role' (Moxnes and Moxnes 2016: 16). This facilitates not only greater cultural understanding, but also potentially change if required—by providing increased awareness of the 'reality' of identity at work. This moment of sense-making could facilitate a change in image—one that can serve as a 'catalyst for reflection and redrafting' (Weick 2005: 416) regarding how the FM discipline defines itself.

In this way, the research answers the call for traditional FM to 'escape the role as the cost reducing administrator of the physical environment and develop a future FM that contributes to creative environments' (Alexander and Price 2012: 62), fashioning communities where identity and meaning are interlinked with 'value creation' as opposed merely to providing 'value for money'. As Endrissat et al. (2015) argue, in the associated creation of service folklore, this transports 'enchantment' towards becoming part of a business model (see also Solnet and Kandampully 2008). Similarly, as Browning and Morris (2012: 1) argue, those who know how to position themselves within an organizational narrative—crucially how to reframe 'stories of life in the workplace'—are at an advantage, with a focus on elements such as project management, strategic planning, and organizational culture.

Finally, as this research shows, three distinct but intertwined professional divisions of FM exist, with the use of the artificial folklore framework providing 'a system of uniting that would otherwise be fragmented' (Smith and Simmons 1983: 377). The different streams of FM are captured in their own characters, plots, and tales, joined culturally in their storylines, and united professionally under the umbrella of fairytales: 'Narratives are . . . seen as central to building community meaning . . . The performance of stories is a key part of members' sense-making and provides a means to allow them to supplant individual memories with institutional memory' (Cunliffe et al. 2004: 265). In so doing, this research contributes to practice as a possible form of workplace learning; with the process of becoming a storyteller—which all the participants were asked to undertake—representing an individual and personal learning experience.

9.3 Limitations and Constraints

> One of the major weaknesses of the logical positivist movement was a tendency on their part to dismiss poetic and metaphorical language as meaningless utterances.
>
> **Knowles and Cole (2008: 9)**

The power of fictional narratives to emplot a discipline incorporates the ability to 'transform itself from its past and present into its future' (Küpers et al. 2013: 84). The tales created here reflect a world created by different service streams within a secondary industry—the plots and characters that professionals form in constructing their everyday corporate realities. However, it must be recognized that there are several limitations and constraints inherent within such a research approach which need to be highlighted. As such, some potential shortcomings and restrictions of the research will now be analysed, with this discussion also including details of mitigating actions implemented in the spirit of remediation.

With the concept of narrative continuing to be 'elusive, contested and indeterminate' (Riessman 2008: 183), it follows that this study can be open to similar criticism. This censure is heightened by recourse in the investigation to fictional allegory, represented by focus on fairy stories underpinned by the material evidence, rather than the more traditional focus on factual lived-experience tales alone. The fictional approach emphasized and utilized in this research could therefore be seen as a limitation, with the argument that it represents primarily a work of 'fiction' and with far fewer lessons to offer for 'reality'. In response, however, we would argue that in this methodology fiction essentially plays a mediating role—a vehicle for generating projective meanings. Indeed, care was taken to address and explain this issue through providing a clear and unambiguous participant brief—one which indicated how the tales created and analysed were to be positioned primarily in terms of metaphor, fable, and allegory.

Linked to this issue of the use of fiction in creating symbolic organizational accounts is the use of ghostwriting in the methodology, that is, the end product of the artificial folklore approach described. This is potentially open to criticism for promoting an extremely partial process—one that is too personal and subjective to be of genuine 'scientific' worth. This point is highlighted by Rhodes and Brown (2005a: 469) in their analysis of the use of fiction in social science, who defend against the claim that 'where research is made up it therefore cannot be trusted'. Although this argument is rejected in this study—due to the detailed methodology developed; one that allowed narrative organizational tales to be constructed—it nevertheless touches upon the issue of validity. Riessman (2008) indicates that narrative studies need to assume validity on two fronts: both in the story told by the participant and the story then told by the researcher.

In terms of the story told by the participant, there is the accepted limitation that this is subjective. However, accessing this subjectivity was the epistemological aim of the research and thus completely in line with its ideographic philosophy. The story told by the researcher meanwhile is visible in this study in two distinct mechanisms. Firstly, as with all qualitative research, there is an accepted limitation that emergent themes, such as identity and image, are fluid, evolving, and therefore not fixed in a form that the participant may be able to express concretely at the moment of inquiry: 'It is unlikely that researchers can access and understand their constitutive assumptions, which are struggling and emerging and hence being realized from day to day' (Johnson and Duberley 2003: 1281). Secondly, the approach of ghostwriting is the inclusion of the voice of the researcher as opposed to the participants themselves. Whilst this could be seen as diluting the analysis, it is positioned precisely to highlight the role of the researcher whose direct voice is often omitted, making this element of the methodology more transparent.

The creation of artificial folklore has been previously discussed, but it remains that the 'artificial' element relates to the absence of an ethnographic approach (McCabe 2013). The tales as collected were thus not 'encountered in their natural states but . . . presented and performed for the benefit of an outsider' (Gabriel and Griffiths 2004: 116). At collection, the tales were simultaneously created into an oral form and then turned into a literary fairytale in the production of this analysis. The artificial folklore framework removed the personal identity of the storyteller in line with the guarantee of research anonymity, leading to a situation where tales were 'removed from their original contexts and re-framed in a literary sphere, where they become self-referential and purely aesthetic items' (Carrassi 2016: 71). However, this potential limitation was mitigated by the inclusion of historical context through the wider FM analysis of Chapter 2, and then at an individual level through developing the lived-experience stories that preceded the re-imaginings. The twin approach of using lived-experience stories in line with fairytale re-imagining also mitigated the risk of the lived-experience stories 'not tallying with the culture' (Jones 1991: 32), with such misalignment potentially occurring due to forces of distortion, fabrication, or through the influence of political dynamics. In the present case, however, the research approach followed was one centred primarily on gleaning individual experiences as opposed to specifying, *tout court*, hypothesized concrete reifications of collective organization.

The concept of using fairytales as an analytical tool can also be seen as a limitation as such an approach necessitates analysis being driven by the relating of one story and one perspective. Fairytales are by their nature univocal. However, this serves to ensure that in the narrative analysis process it is obvious and transparent which point of view is being represented (see Toomeos-Orglaan 2013)—in this case a perspective indicative of a wider occupational collective: 'Given our inherent social need for membership and belonging, our individual life story is often embedded in the story of the community from which we derive our identity' (Yang 2013: 148).

There is also a potential limitation in the timing and length of the study. Presently, these tales provide a snapshot of the FM industry in a period when, among other issues, the United Kingdom is experiencing the effects of global recession. As such, it is a matter of conjecture as to whether these collective images will be temporally sustained (Küpers et al. 2013). Arguably a longitudinal study would allow the evolution of tales, where present, to be charted, and the longevity of reoccurring tales to be identified in terms of the practices of the discipline. This is particularly salient in confirming a state (or otherwise) of disciplinary metamorphosis and identifying the quality and phases of organizational liminality.

The omission of Disney references and imagery could also be considered as a limitation, due to their media universality and dominance in this field. However, it was noted that Disney fairytales are often criticized as being too 'instrumentalized and commercialized' (Zipes 1979: 2) in terms of providing generic historical reference points. Therefore, aligning the method with the more traditional fairytale sources of the Grimm Brothers, Charles Perrault, and Ladybird, we felt, allowed for a more traditional balance of familiarity and authenticity to be struck.

In terms of the implementation of the planned methodology of artificial folklore, the analytical approaches of deconstruction, collaboration, and individual story creation, were, on reflection, not all as successful as envisaged. The deconstruction exercise proved very useful in terms of allowing the constrictions of one allocated story to be highlighted. Here, the limitation of the approach lay in the risk that participants were making their reality fit the storyline of the assigned tale, resulting in a scenario where the 'organizational narratives then become ingredients in the researchers own agendas' (Gabriel 2000: 151). However, the dearth of take-up for the possibility to opt-out of this process—for participants who did not feel sufficient alignment with the story—suggested a general sample affinity with the tale and exercise presented.

The collaboration phases of planned data collection, however, was not as successful as hoped. This was primarily through a failure to achieve sufficient numbers to run the process in the manner foreseen; with the reasons behind this being analysed in Section 5.2. Nevertheless, the effects of this negative outcome for the study were hopefully balanced by achieving larger numbers of volunteers for the individual construction method.

And in terms of the individual construction approach, it can be noted that only the dominant fairytales were analysed in each field, thus discounting the tales of *Little Red Riding Hood*, *Snow White*, *Rapunzel*, *Puss in Boots*, and *Beauty and the Beast*. This, however, was in line with the original research aims which stipulated concentrating on 'reoccurring motifs' rather than 'omissions' (see Hallet and Karasek 1991), a point discussed further in the next section.

9.4 Further Research

This study was innovative, in that a new research methodology was developed to explore the emergent identity of a secondary business service, FM. This has been primarily through assessment of the storied spaces and symbolic archetypes of FM's constituent professional divisions. In being an innovative methodology, however, there is arguably considerable scope for conducting further research into the associated theoretical premises and analytical practices of the research approach.

A potential starting point for such reflection is recognition that the focal research concept, identity, is not static but dynamic (Clegg et al. 2007). As such, developing the framework of artificial folklore to allow for practitioners to cast and recast central characters *over time* could be methodologically beneficial. This could be valuable, for example, in relation to charting evolving elements of the liminal state discussed earlier, and specifically the nature of professional movements therein. In this regard, a longitudinal approach may be recommended—one which could support the use of Jungian archetypes to reveal 'strong plots' (Czarniawska and Rhodes 2006) in organizational and institutional dramas. A longitudinal approach could allow insight into relative movement or stativity, whilst sustaining a constant narrative and thus strengthening the validity of the results presented.

Linked to the last discussion is consideration of futurology issues—in recognition that stories can be used to predict future arrangements of organizational behaviour (Rhodes and Brown 2005b). Although this is not the focus of the present study, fairytales can potentially be used to this end to track the evolution and prospects of aspects of a culture (see Zipes 2011), with the result that this methodological framework could be applied to the temporal tracing of an organization's structures and processes, and specifically the development of new functional disciplines.

Turning from temporal to spatial issues, as noted this study was conducted exclusively with participants in and from the United Kingdom. As such, a future recommendation would be to repeat the process with FM professionals, functions, and divisions further afield—across mainland Europe, and then possibly beyond, to allow a more global picture of the industry to emerge. This is supported by the evidence of several volunteers coming forward from mainland Europe to take part in the research when the initial call was placed on social media.

Another recommendation would be further utilization of the methodology by, on the one hand, targeting different hierarchical levels of FM (and its end-users), and on the other, by application to additional secondary services, such as HRM, IT, and project management. This could serve not only to test the applicability and robustness of the model, but also to capture the storied spaces and symbolic archetypes present in a range of business functions, organizational forms, and industrial sectors. As Gabriel and Griffiths (2004: 123) argue, rather than merely 'providing access to human universals', folklore is an especially appropriate method for 'opening many windows into the idiosyncrasies of each organization'. The narrative framework developed in this study is also flexible in allowing it to be applied to wider organizational issues such as emotion, power, and gender, as well as the sense-making and learning issues discussed above (see Rhodes and Brown 2005b). It is also possible to develop an argument for applying the approach to researching areas of FM reflecting concepts of resistance and irrationality in the workplace (see McCabe 2014, McCabe et al. 2019).

In this study, fairytales have been employed as the main driver of analysis, but this represents just one stream of inquiry within the broader genre of folklore research, as discussed in Chapter 3. Utilizing a folkloric approach more widely within management and organizational studies would, we feel, be beneficial for the field. The folklore literature can be generally sub-divided into the following forms of inquiry: 'research on lore about artefacts and materials, verbal lore such as storytelling, and ritual or customary lore' (Neuman 1999: 444). Modifying the artificial folklore framework to include, for example, issues of the personalization of workspaces, would potentially offer a greater focus on physical objects in relation to social analysis. This may help capture stories reflecting 'socio-material' (Orlikowski 2007, 2010) and 'actor-network' (Alcadipani and Hassard 2010, Law and Hassard 1999) issues and effects in the analysis of organizational functions, customs, and practices. This would potentially support a greater emphasis on the analysis of power in terms of using the framework to illustrate organizational issues related to exercising control and ownership in the FM 'kingdom' (McGee and Pettit 2020).

In the spirit of developing an innovative approach for organizational research, a range of folklore-related methods were explored before the one presented here—artificial folklore—was constructed. As such, another recommendation for further research would be to revisit these early options and explore in greater detail their possibilities for refining the present research approach. For instance, the various story plots and character functions identified by Propp (1968) were considered, with these presenting symbolic references for the development of an analytical approach. The collaborative approach (detailed earlier) could potentially be enhanced in respect of one group of participants being allocated stories generated by a previous group. This could see Propp's '31 functions' either reframed or discarded in order to create a series of tales that the group might now accept more readily. This could function to guide the creation of artificial folklore by capturing individual 'proto stories', or the 'simple pieces of information' that 'carry the seed of a narrative' (Gabriel

1995: 495)—a process potentially allowing new meanings and explanations to be charted.

9.5 Summary

Mirror, mirror on the wall what's the meaning of it all?

As noted in Chapter 1, the overarching aim of developing this study was to produce a new and innovative methodology for organizational research—one taking recourse to an arts-based approach (Rippen 2015, Goulding et al. 2018) to explore, through folkloric analysis, the storied spaces and symbolic archetypes of secondary business services. FM was chosen as a salient example of a 'secondary' service, due to the lack of any generalized agreement regarding what the discipline actually represents (see Drion et al. 2012). This situation has often seen FM perceived in the corporate world variously as a non-core, peripheral, and functionally ambiguous service (Becker 1990)—an organizational netherworld—one aligned increasingly to the potential for outsourcing. This corporate characterizing of FM was in turn contextualized in respect of the discipline frequently operating within a wider context of increasingly fragmented and blurry boundaries in many industrial sectors (Lewis and Kelemen 2002), and notably in a putative state of relentless organizational change. It was with this context in mind, that the present study of FM was prosecuted—a study whose research aims were mainly directed to unlocking narrative patterns of explanation within FM's storied spaces; uncovering the main associated characters and archetypes serving to explain the culture and identity of the discipline, and in so doing, creating an artificial folklore framework and methodology for conducting research in this organizational setting.

Chapter 2 then explored the literature within FM, establishing the three principal 'arenas' of the FM function—people, place, and process (Thompson 1990)—and applying this model to clarification and description of the three delivery mechanisms of consultant, service provider, and in-house FM. Unlike previous research, which had largely focused on occupational roles, industrial contracts, and material concerns of the built environment, the analysis emerging from the present study indicated the key importance of understanding the organizational lifeworlds of practitioners in making sense of the FM 'story'—one that was to lay heavy emphasis on the fluidity of work roles and unconventional organizational issues of professional liminality and service enchantment. The analysis indicated a variety of inherent storied spaces, with no cohesive internal narrative obvious. Instead, the study began to illustrate how facilities managers variously negotiate and reflect upon their positions in different delivery mechanisms, organizational forms, and industrial sector types. This supported the need to investigate FM through acknowledging the multifarious roles of archetypes, rituals, and narratives; factors undergirding development of a novel and bespoke methodology—artificial folklore—for capturing the many symbolic representations of the service (Hassard and Holliday 1998).

This latter line of analysis was expanded in Chapter 3, which primarily related the familiar concept of organizational storytelling to the longstanding tradition of folklore. This association revealed the beginnings of a research approach directed at understanding how 'literary narrative accommodates the unpredictable side of life as well as purposeful behaviour—the fragmented and the coherent at the same time' (Negash 2004 as cited in Watson 2011: 398). This exploration focused on the vehicle of fairytales as a means of explicating issues of professional liminality highlighted in Chapter 2, with this in turn facilitating—in alignment with research aim two—a more robust investigation into issues of narrative archetypes.

Together, these three initial chapters provided the foundations for the research approach advanced within Chapter 4. This represented the planned creation of an artificial folklore framework—a methodology designed to meet the final research aim. In so doing, the discussion details the multiplicity of inputs to the methodology, and how it ultimately represents a combination of methods across storytelling and folklore research. From these foundations, the initial research exercise saw deconstruction of a single tale to allow the story arc and central characters important to the FM practitioner to be identified, an exercise which was supported subsequently by an individual lived-practice analysis and a re-imagining session, where the fairytale chosen was left to the practitioners to decide.

This approach allowed not only the casting of the FM practitioners to be identified within the safety of poetic licence, but also the casting of the different roles within the wider organizational kingdom (see Chapter 5). The main story arcs and archetypes from each FM division were then analysed, with the main motifs forming a ghostwritten fairytale for each one. These three stories illustrate the differences within the discipline, but also the similarities, which formed the key findings explained in Chapter 6 (on the principal tales) and Chapter 7 (on storied spaces).

In Chapters 8 and 9, we reflected on how this research is distinctive in many ways—notably through making reference to the tradition of folklore and merging its historical genres with those of a modern organizational storytelling approach, thus utilizing an 'old skill in a new context' (Snowden 1999: 30). The formation of artificial folklore allows creative insight into this secondary service; which acts functionally as the 'necessary evil' of organization—the *shadow* archetype; manifested as the 'rejected aspects of what the rational organization would like to make itself to be like' (Kociatkiewicz and Kostera 2014: 12). Internally, application of the concept of liminality provides insight into, for example, qualities of ambiguity, states of transition, and rites of passage identified in the research; notably FM's ability to metamorphose, and seek permanence, despite its negative narrative casting. This reflects the quality of FM's identity being eternally in process, in the making, rather than a substantive identity 'product'. In other words, it is 'better understood in terms of becoming rather than being' (Sveningsson and Alvesson 2003: 1165). In this manner, and within this fluctuating organizational landscape, artificial folklore can offer a mechanism for creating 'alternative worlds that put the actual one in a new light' (Bruner 2002: 10); thus, allowing different storied spaces of FM to be perceived.

In sum, this research has illustrated that FM, as an example of a secondary service, does have a mutually shared mythology, a fact that would go some way to explaining why it has persisted as an industry for nearly half a century despite lacking a stable identity. A myriad of contractual and delivery mechanisms representative of this non-core practice is visible and has been described and analysed, albeit silhouetted against a backdrop of frequently shifting and unstable organizational boundaries. FM and its roles are essentially fluid and multi-faceted—a reflection of the current position of secondary services within the 'lifescapes' (Styhre 2017) of modern work organizations. This study hopefully provides some anchor points to help understand this lifeworld of FM and the use of folklore to reveal in organizational terms 'what is expressed and what is denied, what is discovered and what is rejected, to form a picture of the possible world to which Sleeping Beauty, say, will be waking up' (Warner 1994: 418).

APPENDIX 1

Basic Plot Typologies

Aristotle Typology:	**Comic**	**Tragic**	**Romantic**	**Epic**
Booker Typology:	**Comedy**	**Tragedy**	**Rebirth**	**a. Overcoming the Monster** **b. Rags to Riches** **c. The Quest** **d. Voyage and Return**
Protagonist	Deserving victim, a fool. The protagonists are destined to be in love, but something is keeping them from being together, which is resolved by the end of the story.	Undeserving hero/ Victim, villain who falls from grace and whose death is a happy ending.	Villain or otherwise unlikable character who redeems him/herself over the course of the story. Love object	Hero—acts selflessly, is inwardly strong, determined.
Other characters	Trickster	Villain, helper	Gift-giver, lover, injured or sick person	Rescue, object, assistant, villain
Plot Focus/ Predicament	An accident, mistake or coincidence generating emotions of mirth or scorn	A crime or accident insult, injury, loss, mistake, repetition, misrecognition which is often not deserved	Love conquers misfortune, recognition, reciprocation	A challenge, mission or sacrifice where the protagonist is a hero and the plot focus is on achievement: a. The protagonist sets out to defeat an antagonistic force which threatens the protagonist and/or protagonist's homeland. b. The poor protagonist acquires things such as power, wealth, and a mate, before losing it all and gaining it back upon growing as a person.

Continued

Continued

Aristotle Typology:	**Comic**	**Tragic**	**Romantic**	**Epic**
Booker Typology:	**Comedy**	**Tragedy**	**Rebirth**	**a. Overcoming the Monster** **b. Rags to Riches** **c. The Quest** **d. Voyage and Return**
				c. The protagonist and some companions set out to acquire an important object or to get to a location, facing many obstacles and temptations along the way. d. The protagonist goes to a strange land and, after overcoming the threats it poses to him/her, returns with nothing but experience
Emotions that are generated	Mirth, aggression, scorn	Sorrow, pity, fear, anger, pathos, redundancies	Love, care, kindness, generosity, gratitude	Pride, admiration, nostalgia
Function in business	Amusement, aggression, scorn, challenge to org. rationality, power, and legitimacy	Catharsis, sorrow, pity, anger, fear, blame, absolve protagonist of responsibility	Compassion, love, kindness, gratitude, generosity, nostalgia. Organizational takeovers and mergers and the comparison to love affairs and marriage (Hirsch and Andrews as cited in Bowles 1991)	Inspiration, pride, nostalgia, envy
	Victimhood		Hero	

Source: adapted from Hatch and Cunliffe (2006: 199), Gabriel (2000: 84), Booker (2004).

APPENDIX 2

Fairytale Selection Process Example Sheet

ATU	Charles Perrault (1628–1703)	Grimm Brothers 1812–1815	Andrew Lang Collections 1889–1913	Ladybird Classics (1867-present)	Philip Pullman 2013	Naomi Mitchison 1936 Angela Carter 1979 Roald Dahl 1982 1989 Neil Gaiman 1999 2006 2014 Terry Pratchett 1991 Michael Faber 2005	FM References	Occurrences	FM Practi-tioner Responses	Total
510A	Cinderella	Cinderella/ Ashchen-puttel	Cinderella	Cinderella	Cinderella/ One Eyes, Two Eyes, Three Eyes	Cinderella Dahl 1982 Terry Pratchett 1991	Wiggins 2010:1 Marmot & Eley 2000: 49 & 61	9	12	21
333	Little Red Riding Hood	Little Red Cap	Little Red Riding Hood	Little Red Riding Hood	Little Red Riding Hood	The Company of Wolves, Wolf-Alice Carter 1979 Little Red Riding Hood and the Wolf Dahl 1982		7	12	19
327		Hansel and Gretel	Hansel and Gretel	Hansel and Gretel	Hansel and Gretel	Hansel and Gretel Gaiman & Mattioli 2014 *Hansel & Gretel*Mitchison 1936Dahl 1989 The Fahrenheit Twins Faber 2005	Carpman & Grant, 2016: 321	9	9	18
709		Little Snow White		Snow White & The Seven Dwarfs	Snow White	The Snow Child: Carter 1979 Snow, Glass, Apples: Gaiman 1999 Snow White and the Seven Dwarfs Dahl 1982		6	12	18

APPENDIX 3

Artificial Folklore Approach

Research Aims	1. Patterns of FM storied space 2. FM archetypes—internal and external 3. Creation of ghostwritten stories		
Service Streams	Consultant Service Provider In-house	Consultant Service Provider In-house	Consultant Service Provider In-house
Stages	Stage One Deconstruction	Stage Two Collaboration Online	Stage Three Individual Interviews
Activity	Network Mapping	Briefing documents issued, participants numbered randomly 1–6	Introduction of researchers' story
	Creation of character grid	Short questionnaire (table 7)	Short questionnaire (table 7)
	Reading of fairytale *Rumpelstiltskin*	First sentence '*Once upon a time there was*' released	Lived-experience story (BNIM influence)
	Assign on characters and explanation	Each participant creates a paragraph, using the mechanism of the story spine (Parkin 2010) with a vote on the introduction and conclusion.	Creation of fairytale using folkloric framework —Ten fairytales identified and aligned to story typologies —Translated into visual booklet —Character map used for reference
	Option to re-imagine the tale	Follow-up reflective interviews	Fairytale framework used to create a FM tale
	Stage Four **Ghostwriting** Dominant fairytales identified, ATU and Booker typology utilized to create a*rtificial folklore* in the form of a ghostwritten fairytale that represents the tale from each service stream.		

Figure A3 *Artificial Folklore* Approach

APPENDIX 4

Biographic Details of Construction Participants (In Order of Collection)

No	Coding	Gender	Age	Tenure (yrs)	Typology	Sector	Delivery	Level	Fairytale
1	C1	Male	36–45	11–15	Consultant	Private	Total FM	Management	*Little Red Riding Hood*
2	IH1	Male	46–55	15+	In-house	Public	Soft FM	Senior Mgmt.	*Puss in Boots*
3	SP1	Male	55+	15+	Service Provider	Private	Hard FM	Senior Mgmt.	*Goldilocks*
4	C2	Male	36–45	15+	Consultant	Private	Total FM	Executive	*Snow White and the Seven Dwarfs*
5	IH2	Male	55+	15+	In-house	Public	Hard FM	Senior Mgmt.	*Cinderella*
6	IH3	Male	46–55	15+	In-house	Public	Soft FM	Executive	*Beauty and the Beast*
7	IH4	Male	46–55	15+	In-house	Private	Soft FM	Executive	*Cinderella*
8	C4	Female	36–45	<5	Consultant	Public	Total FM	Executive	*Hansel and Gretel*
9	SP2	Male	36–45	5–10	Service Provider	Private	Hard FM	Senior Mgmt.	*Rapunzel*
10	IH5	Female	26–35	11–15	In-house	Private	Total FM	Executive	*Puss In Boots*
11	SP3	Female	36–45	<5	Service Provider	Public	Hard FM	Management	*Goldilocks*
12	IH6	Female	36–45	15+	In-house	Public	Soft FM	Senior Mgmt.	*Cinderella*
13	SP4	Female	46–55	15+	Service Provider	Private	Soft FM	Senior Mgmt.	*Cinderella*
14	C4	Male	36–45	5–10	Consultant	Private	Total FM	Senior Mgmt.	*Hansel and Gretel*

Continued

Continued

No	Coding	Gender	Age	Tenure (yrs)	Typology	Sector	Delivery	Level	Fairytale
15	IH7	Female	36–45	15–10	In-house	Public	Soft FM	Senior Mgmt.	*Cinderella*
16	SP5	Female	46–55	<5	Service Provider	Private	Total FM	Management	*Cinderella*
17	IH8	Female	46–55	11–15	In-house	Public	Soft FM	Senior Mgmt.	*Cinderella*
18	SP6	Male	26–35	<5	Service Provider	Private	Total FM	Management	*Goldilocks*
19	SP7	Male	46–55	15+	Service Provider	Private	Total FM	Director	*Little Red Riding Hood*

Glossary

Allegory Literary device in the form of an extended metaphor—used to reveal hidden or parallel meanings embedded in a story. Allegories are coherent stories in their own right and thus represent effective mechanisms for illustrating complex ideas and concepts (FutureLearn.com).

Alterity Anthropological term indicating a state of being *other*—being different, dwelling in otherness, etc. (Czarniawska and Rhodes 2006).

Archetype In Jungian psychology, an idea or mode of thought inherited from the past through the *collective unconscious.* Visible in folklore through repeating images, motifs, and symbols of fairytales (Ashliman 2004: 180).

Artificial folklore Methodology using multiple story-creating techniques to provide the foundation for the creation of a new tale.

ATU Arne-Thompson-Uther index: a classification system for folk tales developed in 1910 and enlarged in 1928 and 1961. Acts as a 'single classification system by which culturally distinct variants are grouped together according to a common reference number' (Zipes 2000: 1).

Discourse analysis Approach to the analysis of written, vocal, or sign language use, or any significant semiotic event that recognizes there is not one consistent definition of discourse (Saunders et al 2016).

Emic Researching the organization from 'within' methodologically (Jones et al. 1988, Bronner 2017).

Etic Researching the organization from 'without' methodologically (Jones et al. 1988, Bronner 2017).

Fairytale Category of oral folk tale and a genre of prose literature. Specifically contains an element of magic within the narrative (Zipes 2000). Also referred to as *wonder tales.*

Fakelore 'A synthetic product claiming to be authentic oral tradition but actually tailored for mass edification' (Dorson 1963: 94).

Fantasy Denotes anything that is not realist prose and as such can be considered an umbrella term (Zipes 2000). Interchangeable with fairytales in this study.

FM Facilities management/facility management. Organizational support functions defined in this study as three main modes of delivery: consulting provision—consultants working independently on short-term contracts; service provision—firms supplying a range of contracted services over various timescales; and in-house provision—dedicated function of an organization providing support services.

FMers Facility/Facilities managers.

Folk The majority of people in different societies who define themselves through communal associations, whether they interact in groups of long or short duration (Zipes 2005: 12).

Folklife Commonly implies the 'ways the group works together, and the device deployed by the group in carrying out that work... their conjunction perhaps is more important than the distinction' (Abrahams 1978: 162, italics original).

Folklore Collective noun for all information that humans preserve and transmit by personal demonstration and oral communication (Ashliman 2004: 190).

Folktale 'An oral narrative form cultivated by non-literate and literate people to express the manner in which they perceived and perceive nature and their social order' (Zipes 1979: 7).

Legend 'A single event that often seems to transcend natural laws' (Trubshaw 2003: xvi).

Literary fairytale Dating back to the seventeenth century, a work of fiction created by a single, recognized author to imitate—in form, style, and function—the traditional oral magic tales of folklore (Ashliman 2004).

Magical realism 'A combination of the fantastic and the realistic, specifically informed by a narrative tone of banal response to fantastic elements, treating them as equally real to those that are apparently more realistic . . . ordinary events are treated as if they were fantastic and extraordinary events are treated as if they are entirely ordinary' (Stoddart 2007: 35).

Motif Smallest unit within a narrative—basis for much literary and ethnological research (Uther 2004); striking if repeated and becoming part of a tradition.

Myth A 'sacred narrative' explaining how the world and humankind came to be in their present form (Dundes 1997a, 1997b).

Narrative A spoken or written account of connected events—a communicative system containing smaller elements, such as stories (Gabriel 2004).

Phantasmagoria 'A precursor to cinema, phantasmagoria is able to make images move, thereby creating a more effective illusion . . . juxtaposing old-fashioned superstition with new technology' (Buchanan 2010: 366).

Sense-making 'A developing set of ideas with explanatory possibilities, rather than . . . a body of knowledge' (Weick 1995: xi).

Story An emotionally laden symbolic web of plots and characters that contributes to a wider narrative (Gabriel 2004).

Symbolism A process requiring—consciously or unconsciously—an association to create meaning and significance (Pondy et al 1983).

Tale See 'story' above.

Thick description Context of an experience. States the intentions and meaning that organized the experience and reveals experience as a process. Contrasts to thin description, which reports facts independent of intention or circumstances (Denzin and Lincoln 1994: 505).

Wonder tale Few fairytales contain actual reference to fairies, and so many academics refer instead to the supernatural element that defines them, hence deploying the term 'wonder tale' (Trubshaw 2002).

Bibliography

Abma, T. A. 2003. 'Learning by telling: storytelling workshops as an organizational learning intervention'. *Management Learning* 34: 221–240.

Abolafia, M. Y. 2010. 'Narrative construction as sense-making: how central bank thinks'. *Organizational Studies* 31(3): 349–367.

Abrahams, R. D. 1978. 'Towards a sociological theory of folklore: performing services'. *Western Folklore* 37(3): 161–184.

Adams, T. E. 2008. 'A review of narrative ethics'. *Qualitative Inquiry* 14(2): 75–194.

Alcadipani, R., and Hassard, J. 2010. 'Actor-Network theory, organizations and critique: towards a politics of organizing'. *Organization* 17(4): 419–435.

Alexander, K. 1992. 'Facilities management practice'. *Facilities* 10(5): 11–18.

Alexander, K. 2003. 'A strategy for facilities management'. *Facilities* 21(11/12): 269–274.

Alexander, K., and Price I. (eds) (2012). *Managing Organizational Ecologies: Space, Management and Organizations.* London: Routledge.

Alvesson, M., and Skoldberg, K. 2000. *Reflexive Methodology.* London: Sage.

Amaratunga, D., and Baldry, D. 2002. 'Moving from performance measurement to performance management'. *Facilities* 20(5/6): 217–223.

Argyris, C. 2008. *Teaching Smart People How to Learn.* Cambridge, MA: Harvard Business Press.

Arnoux, P., and Soto-Andrade, J. 2019. 'From concrete to abstract and back: metaphor and representation'. Eleventh Congress of the European Society for Research in Mathematics Education, Utrecht University, Netherlands. February.

Ashliman, D. L. 1987. *A Guide to Folktale in the English Language—Based on the Aarne-Thompson Classification System.* New York: Greenwood Press.

Ashliman, D. L. 2004. *Folk and Fairy Tales: A Handbook.* Westport: Greenwood Press.

Atkin, B., and Bildsten, L., 2017 'A future for facility management'. *Construction Innovation* 17(2): 116–124.

Atkin, B., and Brooks, A. 2021. *Total Facilities Management*, fifth edition, London: Wiley-Blackwell.

Bærenholdt, J. O. 2016. 'Experiencing the enchantment of place and mobility'. *Journal of Consumer Care* 16: 393–411.

Bainbridge, M., and Finch, E. F. 2009. 'Getting the attention the facilities manager deserves'. *Facilities* 27(7/8): 277–290.

Baldry, C., and Hallier, J. 2010. 'Welcome to the house of fun: work space and social identity'. *Economic and Industrial Democracy* 31(1): 150–172.

Barnes, V., and Newton, L. 2019. 'Symbolism in bank marketing and architecture: the headquarters of National Provincial Bank of England'. *Management and Organizational History* 14(3): 213–244.

Barrett, P. 2000. 'Achieving strategic facilities management through strong relationships'. *Facilities* 18(10/11/12): 421–426.

Baruch, Y. 2009. 'Once upon a time, there was an organization: organizational stories as antitheses to fairytales'. *Journal of Management Inquiry* 18(1): 15–25.

Baskin, K. 2008. 'Storied spaces: the human equivalent of complex adaptive systems'. *E:CO* 10(2): 1–12.

Bate, S. P. 1997. 'Whatever happened to organizational anthropology? A review of the field of organizational ethnography and anthropological studies'. *Human Relations* 50(9): 1147–1174.

Bauman, Z. 1995. *Life in Fragments: Essays in Postmodern Morality*. Oxford: Blackwell.

Bauman, Z. 2000. *Liquid Modernity*. Cambridge: Polity Press.

Bauman, Z. 2007. *Liquid Times: Living in an Age of Uncertainty*. Cambridge: Polity Press.

Baxter, C. M. 2014. 'The challenge of happily ever after: how once upon a time fan-fic fairytales model strategies for ordinary life changes'. Theses and Dissertations Paper 4128. Available at chrome-extension://efaidnbmnnnibpcajpcglclefindmkaj/https://scholarsarchive.byu.edu/cgi/viewcontent.cgi?article=5127&context=etd

Becker, F. 1990. *The Total Workplace: Facilities Management and the Elastic Organization*. New York: Van Nostrand Reinhold.

Beckett, S. 2014. *Revisioning Red Riding Hood around the World: An Anthology of International Retellings*. Michigan: Wayne State University Press.

Beech, N. 2011. 'Liminality and the practice of identity reconstruction'. *Human Relations* 64(2): 285–302.

Ben-Amos, D. 1971. 'Towards a definition of folklore in context'. *Journal of American Folklore* 84: 167–170.

Benjamin, W. 1968. 'The Task of the Translator' (first printed, 1923). In Z. Horn (ed.), *Illuminations*, pp. 69–82. New York: Harcourt Brace Jovanovich.

Benjamin, W. 1973. 'The Storyteller'. In H. Arendt (ed.), *Illuminations*, pp. 83–109. London: Fontana Collins.

Bennet, E. A. 1982. *Meetings with Jung: Conversations Recorded during the Years 1946–1961*. London: Anchor Press.

Bennett, G. 1993. 'Folklore studies and the English rural myth'. *Rural History* 4(1): 77–91.

Bennett, J. 2001. *The Enchantment of Modern Life: Attachments, Crossings, and Ethics*. Princeton, NJ: Princeton University Press.

Benson, S. 2003. *Cycles of Influence: Fiction, Folklore, Theory*. Detroit: Wayne State University Press.

Berger, P. L., and Luckmann, T. 1966. *The Social Construction of Reality: A Treatise in the Sociology of Knowledge*. London: Penguin Books.

Bettelheim, B. 1976. *The Uses of Enchantment: The Meaning and Importance of Fairy Tales*. London: Penguin Books.

Betts, C. 2009. *Charles Perrault: The Complete Fairy Tales*. Oxford: Oxford University Press.

Blank, T. J., and Howard, R. G. (eds) 2013. *Tradition in the Twenty-first Century: Locating the Role of the Past in the Present*. Colorado: Utah State Press.

Bleistein, S. 2017. *Rapid Organizational Change*. London: John Wiley.

Boje, D. 1991a. 'Consulting and change in the storytelling organization'. *Journal of Organizational Change Management* 4(3): 7–17.

Boje, D. 1991b. 'The storytelling organization: a study of story performance in an office-supply firm'. *Administrative Science Quarterly* 36(1): 6–126.

Boje, D. 1994. 'Organizational storytelling; the struggles of pre-modern, modern and post-modern organizational learning discourses'. *Management Learning* 25(3): 433–461.

Boje, D. 1995. 'Stories of the story-telling organization: a postmodern analysis of Disney as Tamara-Land'. *The Academy of Management Journa*l 38(4): 977–1035.

Boje, D. 2008. 'Storytelling in Management Research'. In R. Thorpe and R. Holt (eds), *Sage Dictionary of Qualitative Management Research*, pp. 213–215. Los Angeles: Sage Publications.

Boje, D. 2011. 'Our organizations were never disenchanted: enchantment by design narratives vs enchantment by emergence'. *Journal of Organizational Change* 24(4): 411–426.

Boje, D., and Baskin, K. 2011. *Dance to the Music of Story: Understanding Human Behavior through the Integration of Storytelling and Complexity Thinking.* New York: IAP.

Bolton, G. 2010. *Reflective Practice: Writing and Professional Development*, third edition. London: Sage Publications.

Booker, C. 2004. *The Seven Basic Plots: Why We Tell Stories.* London: Continuum.

Bottigheimer, R. B. 2003. 'The Ultimate Fairytale: Oral Transmission in a Literate World'. In H. Davidson and A. Chaudhri (eds), *A Companion to the Fairytale*, pp. 57–70. Cambridge: Brewer.

Bowles, M. L. 1989. 'Myth, meaning and work organization'. *Organizations Studies* 10(3): 405–421.

Bowles, M. L. 1991. 'The organizational shadow'. *Organizational Studies* (12/13): 378–404.

Boyce, M. 1996. 'Organizational story and storytelling: a critical review'. *Journal of Organizational Change Management* 9(5): 5–26.

Breton, G. 2009. 'From folk-tales to shareholder-tales: semiotics analysis of the annual report'. *Society and Business Review* 4(3): 87–201.

Brewer, J., and Dourish, P. 2008. 'Storied spaces: cultural accounts of mobility, technology and environmental knowing'. *International Journal of Human-Computer Studies* 66: 963–976.

Breure, A., and Van Meel, J. 2003. 'Airport office: facilitating nomadic workers'. *Facilities* 21(7/8): 75–179.

British Institute of Facilities Management (BIFM). 'BIFM Demographics Report'. Accessed 22 April 2019 at http://www.fm-world.co.uk/features/feature-articles/fms-changing-demographics/

British Institute of Facilities Management (BIFM). 9 June 2014. *FM World—The Death of FM.* BIFM publications.

British Standards Institution (2011). BS EN 15221-3:2011: Facility Management Part 3—Guidance on Quality in Facility Management [online]. London: British Standards Institution.

British Standards Institution (2011). BS EN 15221-4:2011: Facility Management Part 4—Taxonomy, Classification and Structures in Facility Management. [online]. London: British Standards Institution

Brodsky, S. L. 2013. *Testifying in Court: Guidelines and Maxims for the Expert.* Washington, DC: American Psychological Association.

Bronner, S. J. 2016. 'Towards a definition of folklore in practice'. *Cultural Analysis* 15(1): 6–27.

Bronner, S. J. 2017. *Folklore: The Basics.* Oxford: Routledge.

Brown, A. D. 2006. 'A narrative approach to collective identities'. *Journal of Management Studies* 43(4): 731–753.

Brown, A. D., and Toyoki, S. 2013. 'Identity work and legitimacy'. *Organizational Studies* 34(7): 875–896.

Browning, L. D., and Morris, G. H. 2012. *Stories of Life in the Workplace: An Open Architecture for Organizational Narratology.* New York: Routledge.

Bruner, J. 2002. *Making Stories: Law, Literature, Life.* Paris: Farrar, Straus and Giroux

Bryman, A. 2004. *The Disneyization of Society.* London: Sage Publications.

BS EN 15221-3. 2011. (Standards New Zealand) Facility Management, Guidance on quality in Facility Management, published 31 January 2012.

Buchanan, I. 2010. *Oxford Dictionary of Critical Theory*, first edition. Oxford: Oxford University Press.

Buckley, A. D., and Kenney, M. C. 1995. *Negotiating Identity: Rhetoric, Metaphor and Social Drama in Northern Ireland.* London: Smithsonian Institution Press.

Burgoyne, J. 2004. 'Critical Realism'. In R. Thorpe and R. Holt (eds), *Sage Dictionary of Qualitative Management Research*, pp. 64–66. Los Angeles: Sage Publications.

Burner, J. 2002. *Making Stories: Law, Literature, Life.* Cambridge, MA: Harvard University Press.

Burrell, G. 1997. *Pandemonium: Towards a Retro-organization Theory.* London: Sage Publications.

Butler, C., Finniear, J., Doherty, A. M., and Hill, S. 2014. 'Exploring identity: a figurative character image-elicitation approach'. *Qualitative Research in Organizations and Management: An International Journal* 9(2): 51–168.

Business Reporter. 2013. 'Why UK's "hidden industry" must promote itself as a global leader', May. Accessed 22 January 2019 at http://business-reporter.co.uk/2013/05/05/facilities-management/

Cabrelli, M. 2010. 'Facilities Organization: The Hidden Hospital'. In E. Haxby, D. Hunter, and SJagger (eds), *An Introduction to Clinical Governance and Patient Safety*, pp. 241–250. Oxford: Oxford University Press.

Cairns, G. 2003. 'Seeking a facilities management philosophy for the changing workplace'. *Facilities* 21(5/6): 95–105.

Cairns, G. 2014. 'Philosophical Contradictions in FM'. In K. Alexander and I. Price (eds), *Managing Organizational Ecologies: Space, Management and Organizations*, pp. 94–104. London: Routledge.

Calvin, D., and McAra, C. (eds) 2011. *Anti-Tales: The Uses of Disenchantment.* Newcastle Upon Tyne: Cambridge Scholars Publishing.

Carpman, J. R., and Grant, M. A. 2016. Designs That Care: Planning Health Facilities for Patients and Visitors, third edition. London: John Wiley and Sons.

Carrassi, V. 2016. 'A broader and deeper idea of fairy tale: reassessing concept, meaning and function of the most debated genre in folk narrative research'. *Folklore* 65: 70–88.

Carter, A. 1977. *The Fairy Tales of Charles Perrault.* London: Penguin Classics.

Carter, A. 1979. *The Bloody Chamber.* London: Vintage Books.

Carter, A. 1983. 'Notes from the Front Line'. In M. Wander (ed.), *Gender and Writing*, pp. 69–77. London: Pandora Press.

Carter, A. (ed.) 2005. *Book of Fairy Tales.* London: Virago Press.

Casey, C. 2004. 'Bureaucracy re-enchanted? Spirit, experts and authority in organizations'. *Organization* 11(1): 59–79.

Cascio, W., and Montealegre, R. 2016. 'How technology is changing work and organization'. *Annual Review of Organizational Psychology and Organizational Behavior* 3(1): 349–375.

Cassell, C., and Bishop, V. 2014. 'Metaphors and sense-making: understanding the taint associated with dirty work'. *Qualitative Research in Organizations and Management: An International Journal* 9(3): 254–269.

Chartered Institute of Personnel and Development, CIPD. 2022. 'How to manage an increase in flexible working requests', December. Accessed 14 December 2022 at https://www.cipd.co.uk/knowledge/fundamentals/relations/flexible-working/managing-flexible-working-requests-guide

Chen, Z. 2015. 'Principles of facilities management'. *Journal of Facilities Management* 13(1): 3.

Ciacchella, N. 2015. *From the Ashes: A Cinder Will Rise.* London: GGP.

Clark, T., and Salaman, G. 1996. 'Telling Tales: Management Consultancy as the Art of Storytelling'. In D. Grant and C. Oswick (eds), *Metaphor and Organizations*, pp. 97–112. London: Sage Publications.

Clarke, A. C. 1968. Letter to the editor, *Science*, 19 January.

Clegg, S. R., Rhodes, C., and Kornberger, M. 2007. 'Desperately seeking legitimacy: organizational identity and emerging industries'. *Organizational Studies* 28(4): 495–513.

Coenen, C., and Läuppi, V. 2014. Analysis of FM research subjects: A Quantitative Investigation of Academic Facility Management Publications between 2006–2012. Conference proceedings, 13th EuroFM Research Symposium June 2014.

Coenen, C., von Felten, D., and Schmid, M. 2010. 'Reputation and public awareness of facilities management—a quantitative study'. *Journal of Facilities Management* 8(4): 256–268.

Cohen, A. P (ed.) 1986. *Symbolising Boundaries: Identity and Diversity in British Cultures.* Manchester: Manchester University Press.

Collins, D. 2013. 'In search of popular management: sense-making, sensegiving and storytelling in the excellence project'. *Culture and Organization* 19(1): 42–61.

Collins, J., and Porras, J. I. 1994. *Built to Last: Successful Habits of Visionary Companies.* New York: Harper Collins Publishing.

Collins, J. 2001 *Good to great: why some companies make the leap and others don' t.* London: Random House Business.

Collis, J., and Hussey, R. 2003. *Business Research: A Practical Guide for Undergraduate and Postgraduate Students.* London: Palgrave McMillan.

Collison, C., and MacKenzie, A. 1999. 'The power of story in organizations'. *Journal of Workplace Learning* 11(1): 38–40.

Cornelissen, J. P. 2008. 'Metaphor'. In R. Thorpe and R. Holt (eds), *The Sage Dictionary of Qualitative Management Research*, pp. 128–129. Los Angeles: Sage Publications.

Cotts, D., Roper, K., and Payant, R. 2009. *The Facilities Management Handbook*, third edition. London: Amacom.

Cowan, B. 1981. 'Walter Benjamin's theory of allegory'. *New German Critique* 22: 9–122.

Cowls, S. 2015. 'The effects of short-term contracts on organizational culture and engagement within a case study FM organization'. Unpublished paper. Sheffield Business School, Service Sector Department, Sheffield Hallam University.

Crupi, D. 2016. 'Organizational culture and FM responsibilities'. *Facilities Management Journal* Available at https://fmj.ifma.org/articles/organizational-culture-and-fm-responsibilities

Cui, Y. Y., and Coenen, C. 2016. 'Relationship value in outsourced FM services—value dimensions and drivers'. *Facilities* 34(1/2): 43–68.

Cunliffe, A. 2008. 'Social Constructionism'. In R. Thorpe and R. Holt (eds), *The Sage Dictionary of Qualitative Management Research*, pp. 200–202. Los Angeles: Sage Publications.

Cunliffe, A. L., Luhman, J. T., and Boje, D. M. 2004. 'Narrative temporality: implications for organizational research'. *Organizational Studies* 25(2): 261–286.

Czarniawska-Joerges, B. 1992. *Exploring Complex Organizations.* London: Sage Publications.

Czarniawska-Joerges, B. 1995. 'Narration or science? Collapsing the division in organization studies'. *Organization* 2(1): 1–33.

Czarniawska-Joerges, B., and Joerges, B. 1988. 'How to control things with words: organizational talk and control'. *Management Communication Quarterly* 2(2): 70–193.

Czarniawska, B. 1997. *Narrating the Organization: Dramas of Institutional Identity.* Chicago: University of Chicago Press.

Czarniawska, B. 1998. *A Narrative Approach to Organizational Studies.* London: Sage Publications.

Czarniawska, B. 1999. *Writing Management: Organizational Theory as a Literary Genre.* Oxford: Oxford University Press.

Czarniawska, B. 2001. 'Narrative, Interviews and Organizations'. In J. Gubrium and J. Holstein (eds), *Handbook of Interview Research: Context and Method*, pp. 733–750. Thousands Oaks: Sage Publications.

Czarniawska, B. 2004. *Narratives in Social Science Research.* London: Sage.

Czarniawska, B. 2004b. 'Forward: A Semiotic Reading of Strong Plots'. In Y. Gabriel (ed.), *Myths, Stories, and Organizations: Premodern Narratives for our Times*, pp. 1–17. Oxford: Oxford University Press.

Czarniawska, B. 2008. 'Alterity/Identity Interplay in Image Construction'. In D. Barry and H. Hansen (eds), *The Sage Handbook of New Approaches in Management*, pp. 49–62. Los Angeles: Sage Publications.

Czarniawska, B. 2014. 'Nomadic work as life-story plot'. *Computer Supported Cooperative Work* 23: 205–221.

Czarniawska, B., and Gagliardi, P. (eds) 2003. Narratives We Organize By. *Advances in Organization Studies*, Volume 11. Amsterdam: John Benjamin Publishing.

Czarniawska, B., and Mazza, C. 2003. 'Consulting as a liminal space'. *Human Relations* 5(6): 267–290.

Czarniawska, B., and Rhodes, C. 2006. 'Strong Plots: Popular Culture in Management Practice and Theory'. In B. Czarniawska and P. Gagliardi (eds), *Management Education and Humanities*, pp. 195–218. Cheltenham: Edward Elgar.

Dahl, R. 1982. *Revolting Rhymes*. London: Jonathan Cape.

Dahl, R. 1989. *Rhyme Stew*. London: Jonathan Cape.

Dale, K., and Burrell, G. 2010. 'All Together Altogether Better: The Ideal of Community in the Spatial Reorganization of the Workplace'. In A. van Marrewijk and D. Yanow (eds), *Organizational Spaces: Rematerializing the Workaday World*, pp. 20–232. Cheltenham: Edward Elgar.

Davis, O. 2012. *Magic: A Very Short Introduction*. Oxford: Oxford University Press.

Dean, P. 1965. *The First Industrial Revolution*. Cambridge: Cambridge University Press.

de Bruijn, H., van Wezel, R., and Wood, R. 2001. 'Lessons and issues for defining facilities management from hospitality management'. *Facilities* 19(13/14): 476–483.

de Caro, F. (ed.) 2013. *Folklore Recycled—Old Traditions in New Contexts*. Mississippi: University Press of Mississippi.

Dégh, L. 1972 'Folk Narrative'. In *Folklore and Folklife: An Introduction* edited by R. Dorson, pp. 53–84. Chicago: University of Chicago Press.

Dégh, L. 1994. *American Folklore and the Mass Media*. Indiana: Indiana University Press.

Delbridge, R., and Sallaz, J. 2015. 'Work: four worlds and ways of seeing'. *Organizational Studies* 36(11): 449–1462.

Denning, S. 2001. *The Springboard: How Storytelling Ignites Action in Knowledge-era Organizations*. Boston, MA: Butterworth-Heinemann.

Denzin, N., and Lincoln, Y. (eds) 1994. *Handbook of Qualitative Research*. London: Sage Publications.

Donoghue, D. 1983. *The Arts without Mystery*. London: Little, Brown and Company.

Dietz, K., and Silverman, L. 2014. *Business Storytelling for Dummies*. New Jersey: John Wiley and Sons.

Donnelly, P., Gabriel, Y., and Ozkazanc-Pan, B. 2013. 'Untold stories of the field and beyond: narrating the chaos'. *Qualitative Research in Organizations and Management: An International Journal* 8(1): 4–15.

Dorson, R. 1963. 'Current folklore theories'. *Current Anthropology* 4(1): 93–112.

Dorson, R. (ed.) 1972. *Folklore and Folklife: An Introduction*. Chicago: University of Chicago Press.

Douglas, C. 1980. *Bleeders Come First*. London: Fontana.

Drake, C. 1969. 'Jungian psychology and its uses in folklore'. *The Journal of American Folklore* 82(324): 22–131.

Drake, D., and Lanahan, B. 2007. *The Story-Driven Organization*. Global Business and Organizational Excellence 26(4): 36–46.

Drion, B., Melissen F. M and Wood, R. 2012. 'Facilities management: lost or regained?' *Facilities* 30(5/6): 254–261.
Driver, M. 2014. 'The stressed subject: lack, empowerment and liberation'. *Organization* 21(1): 90–105.
Drucker, P. F., 1999 'Knowledge-worker productivity: the biggest challenge'. *California Management Review* 41(2): 79–94.
Dundes, A. 1980. *Interpreting Folklore.* Indianapolis: Indiana University Press.
Dundes, A. 1997a. 'Binary opposition in myth: the Propp/Lévi-Strauss debate in retrospect'. *Western Folklore* 56(1): 39–50.
Dundes, A. 1997b. 'The motif-index and tale type index: a critique'. *The Journal of Folklore Research* 34(3): 95–202.
Dundes, A., and Pagter, C. 1987. *When You're up to your Ass in Alligators: More Urban Folklore from the Paperwork Empire.* Detroit: Wayne State University Press.
Dundon, T., and Ryan, P. 2009. 'Interviewing reluctant respondents: strikes, henchmen and Gaelic games'. *Organizational Research Methods* 13(3): 562–581.
Djerassi, C. 1998. 'Ethical discourse by science-in-fiction'. *Nature* 393: 510–511.
Driver, M. 2009 'Struggling with lack: a Lacanian perspective on organizational identity'. *Organization Studies* 30(1): 55–72.
Earle, H. A., 2003 'Building a workplace of choice: Using the work environment to attract and retain top talent'. *Journal of Facilities Management* 2(3): 244–257.
Easterby-Smith, M., Thorpe, R., and Jackson, P. 2012. *Management Research*, fourth edition. London: Sage Publications.
Elvey, R., Hassell, K., and Hall, J. 2013. 'Who do you think you are? Pharmacist's perceptions of their professional identity'. *International Journal of Pharmacy Practice* 21: 322–332.
Emanuel, D. 2014. 'Where next in FM?' Accessed 15 December 2022 at http://www.i-fm.net/comment/where-next-in-the-fm-journey/
Emmel, N. 2008. 'Participatory mapping: an innovative sociological method. Real life methods toolkit 3'. Accessed 15 December 2022 at http://hummedia.manchester.ac.uk/schools/soss/morgancentre/toolkits/2008-07-toolkit-participatory-map.pdf.
Endrissat, N., Islam, G., and Noppeney, C. 2015. 'Enchanting work—new spirits of service work in an organic supermarket'. *Organization Studies* 36(11): 555–1576.
Enninga, T., and van der Lugt, R. 2016. 'The innovation journey and the skipper of the raft: about the role of narratives in innovation project leadership'. *Project Management Journal* 47(2): 3–114.
Exadaktyloas, A. K., and Velmahos, G. C. 2008. 'Emergency medicine and acute care surgery: a modern *Hansel and Gretel* fairytale?' *Emergency Medical Journal* 25: 321–323.
Exworthy, M., Hyde, P., and McDonald-Kuhne, P. 2016. 'Knights and knaves in the English medical profession: the case of clinical excellence awards'. *Journal of Social Policy* 45(1): 83–99.
Faber, M. 2005 *The Fahrenheit Twins.* Edinburgh: Canongate Books Ltd.
Featherstone, P., and Baldry, D. 2000. 'The value of facilities management function in the UK NHS community health-care sector'. *Facilities* 8(7/8): 302–311.
Feldman, M. S. 1991. 'The Meaning of Ambiguity: Learning from Stories and Metaphors'. In P. J. Frost, L. F. Moore, M. R. Louis, C. C. Lundberg, and J. Martin (eds), *Reframing Organizational Culture*, pp. 145–157. London: Sage Publications.
Fell, J. L. 1977. 'Vladimir Propp in Hollywood'. *Film Quarterly* 30(3): 9–28.
Finch, E. 1992. 'Facilities management at the crossroads'. *Property Management* 10(3): 96–205.
Fisher, W. R. 1984. 'Narration as a human communication paradigm: the case of public moral argument'. *Communication Monographs* 51: 1–22.

Fisseni, B., Kurji, A., and Lowe, B. 2014. 'Annotating with Propp's morphology of the folktale: reproducibility and trainability'. *Literary and Linguistic Computing* 1: 1–23.
Ford, J. D. 1999. 'Organizational change as shifting conversations'. *Journal of Organizational Change* 12(6): 480–500.
Forster, N. 1999. 'The role of storytelling in organizational leadership: leadership and organization'. *Development Journal* 20(11): 11–17.
Fox, N. J. 1992. *The Social Meaning of Surgery*. Buckingham: Open University Press.
Fox, N. J. 1993. 'Discourse, organisation and the surgical ward round'. *Sociology of Health and Illness* 15(1): 16–42.
Fraher, A. L. 2017. 'Invisible dirty work: the multiple realities of US airplane pilots' work'. *Culture and Organization* 23(2): 31–148.
Fraser, J. G. 1890/2009. *The Golden Bough: A Study of Magic and Religion*. Oxford: Oxford University Press.
Gabriel, Y. 1991. 'Turning facts into stories and stories into facts: a hermeneutic exploration of organizational folklore'. *Human Relations* 44(8): 857–874.
Gabriel, Y. 1991b. 'On organizational stories and myths: why it is easier to slay a dragon than to kill a myth'. *International Sociology* 6(4): 427–442.
Gabriel, Y. 1995. 'The unmanaged organization: stories, fantasies and subjectivity'. *Organization Studies* 16(3): 477–501.
Gabriel, Y. 1997. 'Meeting God: when organizational members come face to face with the supreme leader'. *Human Relations* 50(4): 315–342.
Gabriel, Y. 1998. 'Same Old Story or Changing Stories? Folkloric, Modern and Postmodern Mutations'. In D. Grant, T. Keenoy, and C. Oswick (eds), *Discourse and Organization*, pp. 84–103. London: Sage Publications.
Gabriel, Y. 2000. *Storytelling in Organizations: Facts, Fictions, and Fantasies*. Oxford: Oxford University Press.
Gabriel, Y. 2003. 'Your home, my exile: boundaries and "otherness" in antiquity and now'. *Organizational Studies* 24(4): 619–623.
Gabriel, Y. (ed.) 2004. *Myths, Stories, and Organizations: Premodern Narratives for our Times*. Oxford: Oxford University Press.
Gabriel, Y. 2005. 'Glass cages and glass palaces: images of organization in image-conscious times'. *Organization* 12(1): 9–27.
Gabriel, Y. 2008. *Organizing Words—A Critical Thesaurus for Social and Organization Studies*. Oxford: Oxford University Press.
Gabriel, Y. 2012. 'Organizations in a state of darkness: towards a theory of organizational miasma'. *Organizational Studies* 33(9): 137–1152.
Gabriel, Y. 2013. 'Surprises: not just the spice of life but the source of knowledge'. *Management* 16: 719–731.
Gabriel, Y. 2014. 'Book review: *Organizations and Archetypes*'. *Organization* 21(4): 579–581.
Gabriel, Y. 2015. Narratives and Stories in Organizational Life. In A. De Fina and A. Georgeakopoulos (eds), *The Handbook of Narrative Analysis*, first edition, pp. 275–292. London: John Wiley and Sons.
Gabriel, Y., and Connell, N. A. D. 2010. 'Co-creating stories: collaborative experiments in storytelling'. *Management Learning* 41(5): 507–523.
Gabriel, Y., and Griffiths, D. S. 2004. 'Stories in Organizational Research'. In C. Cassell and G. Symon (eds), *Essential Guide to Qualitative Methods in Organizational Research*, pp. 114–124. London: Sage Publications.
Gaiman, N. 1999. *Smoke and Mirrors: Short Fictions and Illusions*. London: Headline.
Gaiman, N. 2006. *Fragile Things: Short Fictions and Wonders*. London: HarperCollins.
Gaiman, N. 2015. *Trigger Warning: Short Fictions and Disturbances*. London: Headline.

Gaiman, N. (author) and Mattotti, L. (illustrator) 2014 *Hansel and Gretel*. London: Bloombury Books.
Garcìa-Rosell, J. 2016. 'A discursive perspective on corporate social responsibility education: a story on co-creation exercise'. *Journal of Business Ethics* 154(4): 1019–1032.
Gaughan, P. 2018. 'FM clinic: debate on BIFM's proposed rebrand and application for chartered body status'. *Facilities Management Journal*. Accessed 15 December 2022 at https://www.fmj.co.uk/fm-clinic-debate-on-bifms-proposed-rebrand-and-application-for-chartered-body-status/2/
Geertz, C. 1973. *The Interpretation of Cultures: Selected Essays*. London: HarperCollins.
George, R. A. 1997. 'The centrality in folkloristics of motif and tale type'. *Journal of Folklore Research* 34(3): 203–208.
Gillet, P. 1998. *Vladimir Propp and the Universal Folktale: Recommissioning an Old Paradigm—-Story as Initiation*. New York: Peter Lang.
Goulding, C., Kelemen, M., and Kiyomiya, T. 2018. 'Community-based response to the Japanese tsunami: a bottom-up approach'. *European Journal of Operational Research* 268(3): 887–903.
Goyal, S., and Pitt, M. 2007. 'Determining the roles of innovation management in facilities management'. *Facilities* 25 (1/2): 46–60.
Grantham, J. W. C. 2003. 'The future of work: changing patterns of workforce management and their impact on the workplace'. *Journal of Facilities Management* 2(2): 42–159.
Green, A. 1978. 'Industrial lore: a bibliographic-semantic query'. *Western Folklore* 37(3): 213–244.
Green, A., and Price, I. 2000. 'Whither FM? A Delphi study of the profession and the industry'. *Facilities* 18(7/8): 281–292.
Greene, C., and Myerson, J. 2011. 'Space for thought designing for knowledge workers'. *Facilities* 29(1/2): 9–30.
Greimas, A. 1966. *Sémantique structurale*. Paris: Presse universitaires de France
Hajduk-Nijakowska, J. 2015. 'Towards the new folklore genre theory'. *Slovak Ethnology: Essays/Discussion* 2(63): 61–173.
Hallett, M., and Karasek, B. (eds) 1991. *Folk and Fairy Tales*. Ontario: Broadview Press.
Hallet, M., and Karasek, B. (eds) 2014. *Fairy Tales in Popular Culture*. London: Broadview Press.
Hamilton, L., and McCabe, D. 2016. 'It's just a job: Understanding emption work, de-animalization and the compartmentalization of organized animal slaughter'. *Organization* 23(3): 330–360.
Hancock, P., and Spicer, A. 2011. 'Academic architecture and the constitution of the new model worker'. *Culture and Organization* 17(2): 91–105.
Hanelt, A., Bohnsack, R., Marz, D., and Marante, C. A. 2021. 'A systematic review of the literature on digital transformation: insights and implications for strategy and organizational change'. *Journal of Management Studies* 58(5): 1159–1197.
Hansen, C. D., and Kahnweller, W. M. 1993. 'Storytelling: an instrument for understanding the dynamics of corporate relationships'. *Human Relations* 46: 391–1409.
Hansen, G. 2001. *The Trickster and the Paranormal*. Bloomington: Xlibris.
Hansen, M. 1987. 'Benjamin, cinema and experience: the blue flower in the land of technology'. *New German Critique* 40: 79–224.
Hansen, W. B., and Altman, I. 1976. 'Decorating personal places: a descriptive analysis'. *Environment and Behaviour* 8(4): 491–504.
Harris, C., Ravenswood, K., and Meyers, B. 2013. 'Glass slippers, holy grails and ivory towers: gender and advancement in academia'. *Labour and Industry: A Journal of the Social and Economic Relations at Work* 23(3): 231–244.

Hassall, J. 1904. *The Old Nursey Stories and Rhymes.* London: Blackie and Son.
Hassard, J. 1991. 'Multiple paradigm research in organizations: a case study. *Organization Studies* 12(2): 275–299.
Hassard, J. 1993. *Sociology and Organization Theory: Positivism, Paradigms and Postmodernity.* Cambridge: Cambridge University Press.
Hassard, J. 2012. 'Rethinking the Hawthorne Studies: the Western Electric research in its social, political and historical context'. *Human Relations* 65(11): 431–1461.
Hassard J., and Holliday R. 1998. *Organization/Representation: Work and Organizations in Popular Culture.* London: Sage.
Hassard, J., and Kelemen, M. 2002. 'Production and consumption in organizational knowledge: the case of the paradigms debate'. *Organization* 9(2): 331–355.
Hassard, J., Kelemen, M., and Wolfram Cox, J. 2008. *Disorganization Theory: Explorations in Alternative Organizational Analysis.* London: Routledge.
Hassard, J., McCann, L., and Morris, J. 2012. '"My brilliant career": new organizational forms and changing managerial careers in Japan, UK and USA'. *Journal of Management Studies* 49(3): 571–599.
Hassard, J., and Morris, J. 2020. 'The extensification of managerial work in the digital age: middle managers, spatio-temporal boundaries, and control'. *Human Relations* 75(9): 1647–1678.
Hassard, J., and Wolfram Cox, J. 2013. 'Can sociological paradigms still inform organizational analysis? A paradigm model for post-paradigm times'. *Organization Studies* 34(11): 701–1728.
Hatch, M., and Cunliffe, A. L. 2006. *Organization Theory: Modern, Symbolic, and Postmodern Perspectives*, second edition. Oxford: Oxford University Press.
Hatch, M., and Jones, M. O. 2007. 'Photocopylore at work: aesthetics, collective creativity and the social construction of organizations'. *Studies in Cultures, Organizations and Societies* 3(2): 263–287.
Hatch, M. J., Kostera, M., and Koźmiński, A. K. 2005. *The Three Faces of Leadership: Manager, Artist and Priest.* London: Blackwell Publishing.
Heath, S. 2016. 'Facilities management is about great service, not trying to do everything'. *Insight* 5 May. Accessed 15 December 2022 at http://workplaceinsight.net/facillities-management-is-about-giving-great-service-not-trying-to-do-everything/
Henson, J. 1989. 'The Storyteller' [television series, USA], Henson Associates, 28 April.
Hightower, R., and Highsmith, J. 2013. 'Investigating the facility management professional shortage'. *International Journal of Facility Management* 4(3): 1–5.
Hirst, A. 2011 'Settlers, vagrants and mutual indifference: unintended consequences of hot-desking'. *Journal of Organizational Change Management* 24(6): 767–788.
Hochschild, A. R. 1983. *The Managed Heart: Commercialization of Human Feeling.* London: University of California Press.
Hodder, A. 2020. 'New technology, work and employment in the era of COVID-19: reflecting on legacies of research'. *New Technology, Work and Employment* 35(3): 262–275.
Hodgson, D. E., and Paton, S. 2015. 'Understanding the profession project management: cosmopolitans, locals and identity work'. *International Journal of Project Management* 34(2): 352–364.
Hokanson, B., and Karlson, R. 2013. 'Borderlands: developing character strengths for a knomadic world'. *On the Horizon* 21(2): 7–113.
Holbek, B. 1987. *Interpretation of Fairy Tales.* Academia Scientarum Fenica. Helsinki: FF Communications.
Hood, S.B.C., 2003 'Delivering minimalist workplaces that improve corporate agility'. *Journal of Facilities Management* 2(1): 68–84.

Höpfl, H. 1995. 'Organizational rhetoric and the threat of ambivalence'. *Studies in Culture, Organizations and Society* 1(2): 75–188.

Höpfl, H. 2000. 'On being moved'. *Studies in Culture, Organizations and Societies* 6(1): 5–34.

Hopland, A. O. 2015. 'Can game theory explain poor maintenance of regional government facilities?' *Facilities* 33(3/4): 95–205.

Houlbrook, C. 2018. 'The concealed revealed'. *History Workshop Journal* 85: 195–216.

Howard, D. 2002. 'Management speak in facilities management'. *Critical Quarterly* 44(4): 25–31.

Hoykinpuro, R., and Ropo, A. 2014. 'Visual narratives on organizational space'. *Journal of Organizational Change Management* 27(5): 792–780.

Huemer, L., Becerra, L., and Lunnan, R. 2004. 'Organizational identity and network identification: relating within and beyond imaginary boundaries'. *Scandinavian Journal of Management* 20(1–2): 53–73.

Hui, E. Y. Y, and Tsang, A. H. C. 2004. 'Sourcing strategies of facilities management'. *Journal of Quality in Maintenance Engineering* 10(2): 85–92.

Hummel, R. P. 1991. 'Stories that managers tell: why are they are as valid as science'. *Public Administration Review* 51: 31–41.

Hunia, F. 1978. *Hansel and Gretel*. Loughborough: Ladybird Books.

Hurtado, K. C., Sullivan, K. T., Okamura, P., and Kashiwagi, D. T. 2014. 'A new instructional strategy and delivery system for the facility management professional (FMP) credential'. *International Journal of Facility Management* 5(2): 1–13.

Hyde, P. 2008. 'Working with stories: diverse tales or organizational life'. *Qualitative Research in Organizations and Management: An International Journal* 3(2): 47–158.

Hyde, P., Sparrow, P., Boaden, R., and Harris, C. 2013. 'High performance HRM: NHS employee perspectives'. *Journal of Health Organization and Management* 27(3): 296–311.

Iedema, R., Long, D., and Carroll, K., 2010 'Corridor communication, spatial design and patient safety: enacting and managing complexities'. In van Marrewijk A and Yanow D (eds) *Organizational Spaces: Rematerializing the Workaday World*, pp. 41–58. Edward Elgar Publishing Limited: Cheltenham.

Ikediashi Dubem, I., Ogunlana, O., and Boateng Prince, S. 2014. 'Determinants of outsourcing decision for facilities management (FM) service provision'. *Facilities* 32(9/10): 472–489.

Imabuchi, S., and Ogata, T. 2014. 'Integrating the event generating mechanism in the Propp-based story generation mechanism into the integrated narrative generation system'. *Journal of Robotics, Networks and Artificial Life* 1(2): 64–168.

Institute of Workplace and Facilities Management, IWFM 2021. 'What is workplace and facilities management?' Accessed 15 December 2022 at https://www.iwfm.org.uk/about/what-is-workplace-and-facilities-management.html.

International Facility Management Association, IFMA 2022 'What is FM?' International Facility Management Association. Accessed 15 December 2022 at https://www.ifma.org/about/what-is-facility-management

Interserve, Sheffield Hallam, and i-FM 2013. 'Time for change in facilities management'. Accessed 21 March 2019 at http://www.interserve.com/docs/default-source/Document-List/services/facilities-management/time-for-change-in-facilities-management.pdf

Islam, G., and Zyphur, M. 2007. 'Ways of interacting: the standardisation of communicaiton in medical training'. *Human Relations* 60(5): 769–792.

ISS 2013. ISS Visions: News ways of working—the workplace of the future. White Book, ISS World Service. Accessed 12 April 2019 at http://www.uk.issworld.com/-/media/issworld/uk/Files/Insights/New_Ways_of_Working_white_book_Apr13.pdf?la=en-GB

ISS 2014. 'Perspective on the FM market development'. ISS White paper, September. Accessed 16 December 2022 at http://www.publications.issworld.com/ISS/External/issworld/White_papers/Perspectives_on_the_FM_market_development/

ISS 2016. 'A proud tradition'. Accessed 16 February 2022 at http://www.uk.issworld.com/about-iss/History

Jackson, R. 1981. *Fantasy: The Literature of Subversion*. London: Routledge.

Jeffres, L., Bracken, C., Jian, G., and Casey, M. 2009. 'The impact of third places on community quality of life'. *Applied Research in the Quality of Life* 4(4): 333–345.

Jemielniak, D. 2008. 'Little Johnny and the Wizard of OS: The PC User as a Fool Hero'. In M. Kostera (ed.), *Organizational Olympians: Heroes and Heroines of Organizational Myths*, pp. 80–92. Basingstoke: Palgrave Macmillan.

Jenkins, R. 2012. 'Disenchantment, enchantment, and re-enchantment: Max Weber at the Millennium'. *Mind and Matter* 10(2): 49–168.

Jensen, P. A., Voordt, T., Coenen, C., Von Felton, D., Lindholm, A., Nielson, S. B., Riratanaphong, R., and Pfenninger, M. 2012. 'In search for the added value of FM: what we know and what we need to learn'. *Facilities* 30 (5/6): 99–217.

Jermier, J. M. 1985. "When the Sleeper Wakes": a short story extending themes in radical organizational theory'. *Journal of Management* 11(2): 67–80.

Jeyaraj, J. S. 2019. 'Teaching literatures: a paradigm shift'. *St. Teresa Journal of Humanities and Social Sciences* 5(1): 55–68.

Johnsen, C. G., and Sørensen, B. M. 2015. 'It's capitalism on coke! From temporary to permanent liminality in organizational studies'. *Culture and Organization* 21(40): 321–337.

Johnson, P., and Duberley, J. 2003. 'Reflexivity in management research'. *Journal of Management Studies* 40(5): 279–1303.

Johnson, P., and Duberley, J. 2006. *Understanding Management Research*. London: Sage Publications.

Johnson, G., Scholes, K., and Whittington, R. 2009. *Fundamentals of Strategy*. Harlow: Pearson Education.

Jones, M. J. 1996. 'Studying organizational symbolism: what, how, why?' *Qualitative Research Methods* 39: 1–80.

Jones, M. O. 1990. 'A folklore approach to emotions at work'. *American Behavioural Scientist* 33(3): 278–286.

Jones, M. O. 1991. 'What if stories don't tally with the culture?' *Journal of Organizational Change* 4(3): 27–34.

Jones, M. O. 1991b. 'On Fieldwork, Symbols and Folklore in the Writing of William Foote Whyte'. In P. J. Frost, L. F. Moore, M. R. Louis, C. C. Lundberg, and J. Martin (eds), *Reframing Organizational Culture*, pp. 124–148. London: Sage Publications.

Jones, M. O., Moore, M. D., and Snyder, R. C. (eds) 1988. *Inside Organizations: Understanding the Human Dimension*. London: Sage Publications.

Jung, C. G. 1964. *Man and his Symbols*. New York: Doubleday.

Jung, C. G. 1969. *The Structure and Dynamics of the Psyche*, Volume 8. London: Routledge and Kegan Paul.

Jung, C. G. 1969b. *The Archetypes and the Collective Unconscious*, Volume 9. London: Routledge and Kegan Paul.

Junghans, A., Olof, N., and Olsson, E. 2014. 'Discussion of FM as an academic discipline: to what extent are the general requirements met and what does it mean?' *Facilities* 32(1/2): 1–12.

Kallio, T. J., Kallio, K., and Blomberg, A. J. 2015. 'Physical space, culture and organizational creativity—a longitudinal study'. *Culture and Organizational Creativity* 33(5/6): 389–411.

Kawalek, P. 2007. 'The bubble strategy: a case study of dynamic, defensible processes of change in Salford'. *International Journal of Public Sector Management* 20(3): 178–191.

Kaya, S., Heywood, C. A., Arge, K., Brawn, G., and Alexander, K. 2004. 'Raising facilities management's profile in organizations: developing a world-class framework'. *Journal of Facilities Management* 3(1): 65–82.

Keenoy, T. 1999. 'HRM as hologram: a polemic'. *Journal of Management Studies* 36(1): 1–23.

Kelemen, M. 2000. 'Too much or too little ambiguity: the language of total quality management'. *Journal of Management Studies* 37(4): 483–498.

Kelemen, M., Philips, M., James, D., and Moffat, S. 2017. 'Performing the Legacy of Animative and Iterative Approaches to Co-producing Knowledge'. In K. Facer and K. Pahl (eds), *Valuing Interdisciplinary Collaborative Research: Beyond Impact*, pp. 107–130. Bristol: Policy Press.

Kelemen, M., and Rumens, N. 2012. 'Pragmatism and heterodoxy in organization research: going beyond the quantitative/qualitative divide'. *International Journal of Organizational Analysis* 20(1): 5–12.

Kelemen, M., Rumens, N., and Vo, L. C. 2018. 'Questioning and organization studies'. *Organization Studies* 40(10): 1529–1542.

Klein, J. H. M. Connell, N. A. D., and Meyer, E. 2007. 'Operational research practice as storytelling'. *The Journal of the Operational Research Society* 58(12): 1535–1542.

Knowles, J. G., and Cole, A. L. (eds) 2008. *Handbook of the Arts in Qualitative Research*. London: Sage Publications.

Kociatkiewicz, J., and Kostera, M. 2011. 'Transitional space'. *Tamara: Journal for Critical Organization Enquiry* 9(3/4): 7–9.

Kociatkiewicz, J., and Kostera, M. 2012. 'The good manager: an archetypical quest for morally sustainable leadership'. *Organizational Studies* 33(7): 861–878.

Kociatkiewicz, J., and Kostera, M. 2012b. 'Sherlock Holmes and the adventure of the rational manager: organizational reason and its discontents'. *Scandinavian Journal of Management* 25: 62–172.

Kociatkiewicz, J., and Kostera, M. 2014. 'Into the labyrinth: tales of organizational nomadism'. *Organizational Studies* 36(1): 1–17.

Kociatkiewicz, J., and Kostera, M. 2016. 'Grand plots of management bestsellers: learning from narrative and thematic coherence'. *Management Learning* 47(3): 324–342.

Kostera, M. 2006. 'The narrative collage as research method'. *Storytelling, Self, Society* 2(2): 5–27.

Kostera, M. (ed.) 2008a. *Mythical Inspirations for Organizational Realities*. Basingstoke: Palgrave Macmillan.

Kostera, M. (ed.) 2008b. *Organizational Epics and Sagas: Tales of Organizations*. Basingstoke: Palgrave Macmillan.

Kostera, M. (ed.) 2008c. *Organizational Olympians: Heroes and Heroines of Organizational Myths*. Basingstoke: Palgrave Macmillan.

Kostera, M. 2012. *Organizations and Archetypes*. Cheltenham: Edward Elgar.

Kostera, M., and Oblój, K. 2010. 'Archetypes of rivalry: narrative responses of Polish radio station managers to perceived environmental change'. *Journal of Organizational Change* 23 (5): 564–577.

Kramer, A., and Kramer, K. Z. 2020. 'The potential impact of the Covid-19 pandemic on occupational status, work from home, and occupational mobility'. *Journal of Vocational Behavior* 119: 03442. Available at: https://doi.org/10.1016/j.jvb.2020.103442

Krawczyk-Wasilewska, V. 2006. 'e-Folklore in the age of globalization', *Fabula* 47(3–4): 248–254.

Kress, N. 1993. *Beginnings, Middles, and Ends*. Ohio: Writer's Digest Books.

Küpers, W. 2011. 'Dancing on the limen—embodied and creative inter-places as thresholds of be(coming): phenomenological perspectives on liminality and transitional spaces in organisation and leadership'. *Tamara Journal for Critical Organization Inquiry* 9(3/4): 45–59.

Küpers, W., Mantere, S., and Statler, M. 2013. 'Strategy as storytelling: a phenomenological collaboration'. *Journal of Management Inquiry* 22(1): 83–100.

Kvale, S. 1996. *InterViews: An Introduction to Qualitative Research Interviewing*. Thousand Oaks: Sage Publications.

Lacan, J. (1977). *Écrits*. London: Tavistock Publications.

Lacey, N. 2000. *Narrative and Genre*. London: Macmillan.

Lakoff, G., and Johnson, M. 1980. *Metaphors We Live By*. Chicago: University of Chicago Press.

Landau, D., Drori, I., and Terjesen, S. 2014. 'Multiple legitimacy narratives and planned organizational change'. *Human Relations* 67(11): 1–25.

Lang, A. 1901. *Magic and Religion*. London: Longmans, Green and Co.

Lam, B., Phillips, M., Kelemen, M., Zamenopoulos, T., Moffat, S., and deSousa, S. 2018. 'Design and creative methods as a practice of liminality in community–academic research projects'. *The Design Journal* 21(3): 602–624.

Lambert, A. 1989. 'Corporates identity and facilities management'. *Facilities* 7(12): 7–12.

Larsen, J. L., Elle, M., Hoffman, B., and Munthe-Kaas, P. 2011. 'Urbanising facilities management: the challenges in a creative age'. *Facilities* 29 (1/2): 80–92.

Latham, A. 1999. 'The power of distraction: distraction, tactility and habit in the work of Walter Benjamin'. *Environment and Planning: Society and Space* 17: 451–473.

Lave, J., and Wenger, E. 1991. *Situated Learning: Legitimate Peripheral Participation*. Cambridge: Cambridge University Press.

Law, J. 2009. 'A Dictionary of Business and Management' [online]. Oxford: Oxford University Press. Accessed 16 December 2022 at https://books.google.co.uk/books?id=zR9cBgAAQBAJandprintsec=frontcoverandsource=gbs_ge_summary_randcad=0#v=onepageandqandf=false

Law, J., and Hassard, J. (eds) 1999. *Actor Network Theory and After*. Oxford: Blackwell.

Leach, E. (ed.) 1968. *The Structural Study of Myth and Totemism*. Norfolk: Cox and Wyman.

Leaman, A. 1992. 'Is facilities management a profession?' *Facilities* 10(10): 18–20.

Lefebvre, H. 1991. *The Production of Space*. London: Wiley-Blackwell.

Lévi-Strauss, C. 1955. 'The structural study of myth'. *The Journal of American Folklore* 270(68): 428–444.

Lévi-Strauss, C. 1978. *Myth and Meaning*. New York: Schocken Books.

Lévi-Strauss, C. 1979 *The Origin of Table Manners: Introduction to a Science of Mythology*, Volume 3. New York: Harper and Row.

Lewis, M. W., and Kelemen, M. L. 2002. 'Multiparadigm inquiry: exploring organizational pluralism and paradox'. *Human Relations* 55(2): 251–275.

Lewis, P. L. 2011. 'Storytelling as research/research as storytelling'. *Qualitative Inquiry* 17(6): 505–510.

Lim, W. M. 2021. 'Editorial: history, lessons, and ways forwards from the Covid-19 pandemic'. *International Journal of Quality and Innovation* 2(2): 100–108.

Lindahl, C. 1997. 'Some uses of numbers'. *Journal of Folklore Research* 34(3): 264–273.

Lowry, D. 2017. *The Complete Guide to Facility Management*. Scotts Valley, CA: CreateSpace.

Lucas, M. 2014. 'Nomadic organizations and the experience of journeying: through liminal spaces and organizing places'. *Culture and Organization* 20(3): 1–19.

Lüthi, M. 1976. *Once upon a Time: On the Nature of Fairy Tales*. Bloomington: Indiana University Press.

Lüthi, M. 1991. 'The Fairy-tale Hero: The Image of Man in the Fairy Tale'. In M. Hallett and B. Karasek (eds), *Folk and Fairy Tales*, pp. 314–326. Ontario: Broadview Press.
MacDonald, R., Price, I., and Askham, P. 2009 'Excellent patient environments within acute NHS trusts'. *Journal of Facilities Management* 7(1): 7–23.
MacDonald, C. L., and Sirianni, C. 1996. *Working in the Service Society*. Philadelphia: Temple University Press.
Macleod, D. 2008. *Enchanted Lives, Enchanted Objects: American Women Collectors and the Making of Culture, 1800–1940*. San Francisco: University of California Press.
Manyika, J. 2017. 'Technology, jobs, and the future of work'. McKinsey & Co. Executive Briefing. 24 May.
Markus, T. A., and Cameron, D. 2002. *The Words between the Spaces: Building and Language*. London: Routledge.
Marmot, A., and Eley, J. 2000. *Office Space Planning: Designing for Tomorrow's Workplace*. London: McGraw-Hill Education.
Martin, J., Feldman, M. S., Hatch, M. J., and Sitkin, S. B. 1983. 'The uniqueness paradox in organizational stories'. *Administrative Science Quarterly* 28: 438–453.
Mazumdar, S. 1992. '"Sir, please do not take away my cubicle": the phenomenon of environmental deprivation'. *Environment and Behavior* 24(6): 691–722.
McCabe, D. 2009. 'Strategy-as-power: ambiguity, contradiction and the exercise of power in a UK building society'. *Organization* 17(2): 51–175.
McCabe, D. 2013. 'The tyranny of distance: Kafla and the problem of distance in bureaucratic organizations'. *Organization* 22(1): 58–77.
McCabe, D. 2014. 'Making out and making do: how employees resist and make organizational change work through consent in a UK bank'. *New Technology, Work and Employment* 21(1): 57–71.
McCabe, D. 2016. 'Curiouser and curiouser! Organizations as Wonderland —a metaphorical alternative to the rational model'. *Human Relations* 69(4): 1–29.
McCabe, D., Ciuk, S., and Gilbert, M. 2019. 'There is a crack in everything: an ethnographic study of pragmatic resistance in a manufacturing organization'. *Human Relations* 73(7): 953–980.
McCann, L., Hassard, J. S., Granter, E., and Hyde, P. J. 2015. 'Casting the lean spell: the promotion, dilution and erosion of lean management in the NHS'. *Human Relations* 68(10): 557–1577.
McCarroll, P. 2017. 'The house that facilities management built: a story of identity and metaphor in a secondary service'. *Culture and Organization* 23(4): 291–305.
McGee, R., and Pettit, J. (eds) 2020. *Power, Empowerment and Social Change*. London: Routledge.
McGlathery, J. M. (ed.) 1991. *The Brothers Grimm and Folktale*. Chicago: University of Illinois Press.
McWhinney, W., and Batista, J. 1988. 'How remythologisingcan revitalise organizations'. *Organizational Dynamics* 17(2): 46–58.
Meyerson, D. 1991. 'Normal Ambiguity? A Glimpse of an Occupational Culture'. In P. J. Frost, L. F. Moore, M. R. Louis, C. C. Lundberg, and J. Martin (eds), *Reframing Organizational Culture*, pp. 131–145. London: Sage Publications.
Mehrabian, A., and Russell, J. A. 1974. *An Approach to Environmental Psychology*. Cambridge, MA: Massachusetts Institute of Technology.
Mitchison, N. 1936. *The Fourth Pig*. Princeton, NJ: Princeton University Press.
Mitchell-Ketzes, S. 2003 'Optimising business performance through innovative workplace strategies'. *Journal of Facilities Management* 2(3): 258–275.

Modell, S. 2009. 'In defence of triangulation: a critical realist approach to mixed methods research in management accounting'. *Management Accounting Research* 20(2): 208–221.

Monin, N., and Monin, J. 2005. 'Hijacking the fairytale: genre blurring and allegorical breaching in management literature'. *Organization* 12(4): 511–528.

Moravec, J. W. 2013. 'Knowmad society: the "new" work and education'. *On the Horizon* 21(2): 79–83.

Morgan, G. 1993. *Imaginization: The Art of Creative Management*. London: Sage Publications.

Morgan, G. 1998. *Images of the Organization*. London: Sage Publications.

Morrell, K., and Tuck, P. 2014. 'Governance, tax and folk tales'. *Accounting, Organizations and Society* 39(2): 37–147.

Moxnes, P. 1999a. 'Deep roles: twelve primordial roles of mind and organization'. *Human Relations* 52(11): 427–1444.

Moxnes, P. 1999b. 'Understanding roles: a psychodynamic model for role differentiation in groups'. *Group Dynamics: Theory, Research and Practice* 3(2): 99–113.

Moxnes, P. 2013. 'The hero's dream and otherprimordial patterns of imagery: archetypal influences on organizational fantasies and ideations'. *Journal of Organizational Change Management* 26(4): 638–653.

Moxnes, P., and Moxnes, A. 2016. 'Are we sucked into fairy tale roles? Role archetypes in imagination and organization'. *Organizational Studies* 37(10): 1519–1539.

Mudrak, T., Wagenberg, A. V., and Wubben, E. 2004. 'Assessing the innovative ability of FM teams: a review'. *Facilities* 22(11/12): 290–295.

Mudrak, T., Wagenberg, A. V., and Wubben, E. 2005. 'Innovation process and innovativeness of facility management organizations'. *Facilities* 23(2/3): 3–118.

Myeda, N E., 2014. 'Facilities management: the business enabler'. *Journal of Facilities Management* 12(4): 1–4.

Nenonen, S., Lavy, S., Graça, M., Crepaldi, R., and Sandström, N. 2023. 'Future of the facilities management profession'. IOP Conference Series: Earth and Environmental Science, 1176. https://doi.org/0.1088/1755-1315/1176/1/012008.

Neuman, L. J. 1999. 'Paper, piles and computer files: folklore of information work environments'. *Library Trends* 47(1): 439–469.

Noor, M. N. M., and Pitt, M. 2009. 'A critical review on innovation in facilities management service delivery'. *Facilities* 27(5/6): 211–228.

Nor, N. A. M., Mohammad, A. H., and Alias, B. 2014. 'Faculty management history and evolution'. *International Journal of Facility Management* 5(1): 1–12.

Nollendorfs, C. L. 1983. 'Fairytales for language instruction: poisoned apple or gold from straw?' *Die Unterriehtspraxis/ Teaching German* 16(2): 290–294.

Nutt, B. 2000. 'Four competing futures for facility management'. *Facilities* 18(3/4): 24–132.

O'Briain, D. 2007. 'Presentation'. British Institute of Facilities Management (BIFM) Annual Excellence Awards, available at http://www.bifm.org.uk/bifm/news/4418

O'Connor, E. 2002. 'Storied business: typology, intertextuality and traffic in entrepreneurial narrative'. *The Journal of Business Communication* 39(1): 36–54.

Okhuysen, G., Lepak, D., Ashcraft, K., Labianca, G., Smith, V., and H. Steensma. 2013. 'Theories of work and working today'. *Academy of Management Review* 38(4): 491–502.

Oldenburg, R. 1989. *The Great Good Place: Cafes, Coffee Shops, Community Centres, Beauty Parlours, General Stores, Bars, Hangouts, and How They Get You through the Day*. New York: Paragon House.

Opie, I., and Opie, P. 1980. *The Classic Fairy Tales*. Oxford: Oxford University Press.

Orlikowski W. J. 2007. 'Sociomaterial practices: exploring technology at work'. *Organization Studies* 28: 1435–1448.

Orlikowski, W. 2010. 'The sociomateriality of organizational life: considering technology in management research'. *Cambridge Journal of Economics* 34: 125–141.

Orr, J. E. 1996. *Talking about Machines: Ethnography of a Modern Job*. Ithaca: Cornell University Press.

Orwell, G. 1945/2000. *Animal Farm: A Fairy Story*. London: Penguin Modern Classics.

Oswick, C. 2008. 'Narrative Research'. In R. Thorpe and R. Holt (eds), *The Sage Dictionary of Qualitative Management Research*, pp. 141–142. Los Angeles: Sage Publications.

Park, Y. 1982. 'The function of fiction'. *Philosophy and Phenomenological Research* 42(3): 416–424.

Parkin, M. 2010. *Tales for Change: Using Storytelling to Develop People and Organizations*. London: Kogan Page.

Pässilä, A., Oikarinen, T., and Kallio, A. 2013. 'Creating dialogue by storytelling'. *Journal of Workplace Learning* 25(3): 59–177.

Pathirage, C., Haigh, R., Amaratunga, D., and Baldry, D. 2008. 'Knowledge management practices in facilities organization: a case study'. *Journal of Facilities Management* 6(1): 522–537.

Pennock, G. A. 1930. 'Industrial research at Hawthorne'. *Personnel Journal* 8: 296–313.

Perrault, C. 1697/2020. *The Tales of Mother Goose*. London: Pharos Books.

Petani, F. J., and Mengis, J. 2016. 'In search of lost space: the process of space planning through remembering and history'. *Organization* 23(1): 71–89.

Petit, S., Mougenot, M., and Fleury, P. 2011. 'Stories on research, research on stories'. *Journal of Rural Studies* 27: 394–402.

Peters, T., and Waterman, R. 1982. *In Search of Excellence: Lessons from America's Best-run Companies*. New York: Harper and Row.

Philips, M., and Rippin, A. 2010. 'Howard and the mermaid: abjection and the Starbucks Foundation memoir'. *Organization* 17(4): 481–499.

Phillips, N. 1995. 'Telling organizational tales: on the role of narrative fiction in the study of organizations'. *Organizational Studies* 16(4): 625–649.

Phillips, N., and Zyglidopoulos, S. 1999. 'Learning for foundation: Asimov's psychohistory and limits of organizational theory'. *Organization* 6(4): 591–608.

Pearce, B., Hinds, P., Thomason, B., Altman, H., and Winterstorm, S. 2023. 'Cultivating place identity at work'. *Organizational Dynamics* 52(3): 100997. Available at https://doi.org/10.1016/j.orgdyn.2023.100997

Pondy, L. R., Frost, P. J., Morgan, G., and Dandridge, T. C. (eds) 1983. *Organizational Symbolism*. London: Jai Press.

Pratchett, T. 1991. *Witches Abroad: A Discworld Novel*. London: Victor Gollancz Ltd.

Price, I. 2001. 'Can FM evolve? If not, what future?' *Journal of Facilities Management* 1(1): 59–69.

Price, I. 2003. 'Facility management as an emerging discipline'. In R. Best, C. Langston, and G. deValence (eds), *Workplace Strategies and Facilities Management*, pp. 49–65. Oxford: Butterworth Heinemann.

Price, I. 2004. 'Business-critical FM'. *Facilities* 22(13/14): 353–358.

Price, I. 2007. 'Lean assets: new language for new workplaces'. *California Management Review* 49(2): 2–118.

Price, I. 2011. 'Triple bottom line space: Productivity Paradoxes in FM'. 10th EuroFM Research Symposium: Cracking the Productivity Nut, Vienna 24–25 May. Available at http://shura.shu.ac.uk/id/eprint/3769

Price, I. 2012. 'The selfish signifier—meaning, virulence and transmissibility in a management fashion'. *International Journal of Organizations Analysis* 20(3): 337–348.

Price, I., and Akhlaghi, F. 1999. 'New patterns in facilities management: industry best practice and new organizational theory'. *Facilities* 17(5/6): 59–166.

Price, I., and Fortune, J. 2008. 'Open plan and academe: pre-and post-hoc conversations'. In Proceedings of the W070 conference: health and creative facilities edited by D. Then and E. Finch, pp. 613–620. CIB publications (315).

Price, I., and Shaw, R. 1998. *Shifting the Patterns: Breaching the Memetic Codes of Corporate Performance*. London: Management Books.

Price, C., and Whiteley, A. 2014. 'Corporate culture and employee identity: cooption or commitment through contestation?' *Journal of Change Management* 14(2): 210–235.

Propp, V. 1968. *Morphology of the Folktale* (second edition, first edition 1928). Austin: University of Texas Press.

Puddy, R., Price, I., and Smith L. 2001. 'FM policies and standards as a knowledge management system'. *Facilities* 19(13/14): 504–515.

Pullman, P. 2012. *Grimm Tales for Young and Old*. London: Penguin Books.

Raelin, J. A. 2012. 'The manager as a facilitator of dialogue'. *Organization* 20(6): 818–839.

Ramseook-Munhurruni, P. 2016. 'A critical incident technique investigation of customers' waiting experiences in service encounters'. *Journal of Service Theory and Practice* 26(3): 246–272.

Read, M. 2016. Where there is FM, there is magic. *FM World* article, 26 April. Accessed 16 May 2018 at http://www.fm-world.co.uk/comment/blog/where-there-is-fm-there-is-magic/

Reissman, C. K. 2008. *Narrative Methods for the Human Sciences*. New York: Sage Publications.

Rhodes, C. 2000. 'Ghost-writing research: positioning the researcher in the interview text'. *Qualitative Inquiry* 6(4): 511–525.

Rhodes, C. 2001. *Writing Organizations: (Re)Presentation and Control Narratives of Work*. Amsterdam: John Benjamins Publishing.

Rhodes, C. 2015. 'Writing organization/romancing fictocriticsm'. *Culture and Organization* 21(4): 289–303.

Rhodes, C., and Brown, A. D. 2005a. 'Writing responsibility: narrative fictions and organizational studies'. *Organization* 12(4): 467–491.

Rhodes, C., and Brown, A. D. 2005b. 'Narrative, organizations and research'. *International Journal of Management Review* 7(3): 67–188.

Richardson, L. L. 1992. 'The Consequences of Poetic Representation: Writing the Other, Rewriting the Self'. In C. Ellis and M. G. Flaherty (eds), *Investigating Subjectivity: Research on the Lived Experience*, pp. 125–137. Newbury Park: Sage Publications.

Riessman, C. K. 2008. *Narrative Methods for the Human Sciences*. London: Sage Publications.

Rippin, A. 2011. 'A visit to the "Great Ghaytez's palace": a case study in using a literary genre to explore the effects of corporate architecture'. *Journal of Organizational Change Management* 24(6): 720–732.

Rippin, A. 2011b. 'Ritualised Christmas headgear or "Pass me the tinsel, mother: it's the office party tonight"'. *Organization* 18(6): 823–832.

Rippin, A. 2013. 'Putting the Body Shop in its place: a studio-based investigation into the new sites and sights of organization as experience'. *Organization Studies* 34(10): 551–1562.

Rippin, A. 2015. 'Feminine writing: text as dolls, drag and ventriloquism'. *Gender, Work and Organization* 22(2): 12–128.

Ritzer, G. 1993. *The McDonaldization of Society* (also 1999, 2000, 2005 editions). Thousand Oaks: Pine Forge Press.

Roberts, S. 2015. 'Models of enchantment and the enchantment of models'. *Journal of Business Anthropology* 4(1): 58–164.

Robichaud, D. 2003. 'Narrative Institutions We Organize By'. In B. Czarniawska and P. Gagliardi (eds), *Advances in Organization Studies*, Volume 11, pp. 37–55. Amsterdam: John Benjamin Publishing.

Robinson, L., and Wildersmith, S. M. 2015. 'From rags to splendour: the evolution of *Cinderella* cover illustrations from 1800 to 2014'. *Visual Communications* 15(1): 54–70.

Rotmann, S., Mourik, R., and Goodchild, B. 2015. 'Once Upon a Time . . . How to tell a good energy efficiency story that "sticks"'. In ECEEE 2015 Summer Study, Proceedings: First Fuel Now, 1–6 June, pp. 113–122. Stockholm: ECEEE.

Rosen, M. 1985. 'Breakfast at Spiro's: dramaturgy and dominance'. *Journal of Management* 11(2): 31–48.

Ross, J. P., Intindola, M. L., and Boje, D. M. 2016. 'It was the best of times; it was the worst of times: the expiration of work-life balance'. *Journal of Management Inquiry* 26(2): 202–215.

Rostron, A. 2014. 'Imagination, fiction and real life: revitalising the interview in identity studies'. *Electronic Journal of Business Research Methods* 12(2): 96–102.

Rowling, J. K. 1997. *Harry Potter and the Philosopher's Stone*. London: Bloombury Publishing.

Rowling, J. K. 2007. *Harry Potter and the Deathly Hallows*. London: Bloombury Publishing.

Royal Institute of Chartered Surveyors (RICS) October 2013. Strategic Facilities Management, first edition. Accessed 9 April 2019 at http://www.rics.org/Global/Strategic_Facilities_Management_2nd_edition_PGguidance_2013.pdf

Royal Institute of Chartered Surveyors (RICS) November 2013. Raising the Bar 1: Enhancing the Strategic Role of FM. Accessed 18 March 2021. https://www.rics.org/uk/news-insight/research/research-reports/raising-the-bar-strategic-fm/#III

Saunders, M., Lewis, P., and Thornhill, A. (eds) 2016. *Research Methods for Business Students*, seventh edition. Harlow: Pearson Education.

Schneider, S. C., and Shrivastaval, P. 1988. 'Basics assumptions themes in organizations'. *Human Relations* 41(7): 493–515.

Scott, S., and Lane, V. 2000. 'Fluid, fractured, and distinctive? In search of a definition of organizational identity'. *Academy of Management Review* 25(1): 143–144.

Shadraconis, S. 2013. 'Leaders and heroes: modern-day archetypes'. *Lux: A Journal of Transdisciplinary Writing and Research from Claremont Graduate University* 3(1): 1–13.

Sharp, B. 2013. 'Royal paternalism and the moral economy in the reign of Edward the Second: The response to the great famine'. *The Economic History Review* 66(2): 628–647.

Shearer, A. W. 2004. 'Applying Burke's dramatic pentad to scenarios'. *Futures* 36: 823–835.

Shishkoff, 1976. 'The structure of fairy tales: Propp vs Lévi-Strauss', *Disposito*, 1(3): 271–276. https://monoskop.org/images/1/15/Shishkoff_Serge_1976_The_Structure_of_Fairytales_Propp_vs_Levi-Strauss.pdf

Shockley-Zalabak, P. 2002. 'Protean places: teams across time and space'. *Journal of Applied Communication Research* 30(3): 231–250.

Shokeid, M. 1982. 'Toward an anthropological perspective of fairy tales'. *Sociological Review* 30(2): 223–233.

Shove, E. 1991. 'Letterset zombies?' *Facilities* 9(8): 9–12.

Shrive, T. 2023. 'Is AI the future of FM?' *Facilities Management Journal*, 10 May. Available at: https://monoskop.org/images/1/15/Shishkoff_Serge_1976_The_Structure_of_Fairytales_Propp_vs_Levi-Strauss.pdf

Shull, K. K. (2005). 'Is the magic gone? Weber's "disenchantment of the world" and its implications for art in today's world'. *The Culture Issue* 61. Accessed 12 April 2019 at http://www.nyu.edu/pubs/anamesa/archive/fall_2005_culture/11_shull.pdf

Silverman, D. 2011. *Interpreting Qualitative Data: A Guide to the Principles of Qualitative Research*, fourth edition. London: Sage Publications.

Silverman, D. (ed.) 2021. *Qualitative Research*, fifth edition. London: Sage

Simpson, J., and Roud, S. (eds) 2003. *A Dictionary of English Folklore.* Oxford: Oxford University Press.

Simpson, R., Slutskaya N., Lewis, P., and Höpfl, H. (eds) 2012. *Dirty Work: Concepts and Identities.* London: Palgrave.

Sims, D. 2003. 'Between the milestones: a narrative account of the vulnerability of middle managers' storytelling'. *Human Relations* 56(10): 1195–1211.

Smircich, L. 1983. 'Concepts of culture and organizational analysis'. *Administrative Science Quarterly* 28(3): 339–358.

Smith, K., and Simmons, V. M. 1983. 'A Rumpelstiltskin organization: metaphors on metaphors in field research'. *Administrative Science Quarterly* 28(3): 377–392.

Smith, R., Pederson, S., and Burnett, S. 2014. 'Towards an organizational folklore of policing: the storied nature of policing and the police use of storytelling'. *Folklore* 125: 218–277.

Snowden, D. 1999. 'Storytelling: an old skill in a new context'. *Business Information Review* 16(1): 30–37.

Søderberg, A. 2003. 'Sensegiving and Sense-making in an Integration Process: A Narrative Approach to the Study of an International Acquisition'. In B. Czarniawska and P. Gagliardi (eds), *Narratives We Organize By: Advances in Organization Studies*, Volume 11, pp. 3–37. Amsterdam: John Benjamin Publishing.

Solnet, D., and Kandampully, J. 2008. 'How some service firms have become part of service excellence folklore'. *Managing Service Quality* 18(2): 79–193.

Southgate, V. 1964. *Cinderella: A Ladybird Easy-Reading Book.* Loughborough: Wills and Hepworth.

Southgate, V. 1964b. *Hansel and Gretel: Ladybird Tales.* London: Ladybird Books.

Southgate, V. 1965. *Jack and the Beanstalk: A Ladybird Easy-Reading Book.* Loughborough: Wills and Hepworth.

Southgate, V. 1968. *Rapunzel: A Ladybird Easy-Reading Book.* Loughborough: Wills and Hepworth.

Southgate, V. 1969. *Sleeping Beauty: A Ladybird Easy-Reading Book.* Loughborough: Ladybird Publishing.

Southgate, V. 1969b. *Snow-white and the Seven Dwarfs: A Ladybird Easy-Reading Book.* Loughborough: Wills and Hepworth.

Southgate, V. 1972. *Little Red Riding Hood: A Ladybird Easy-Reading Book.* Loughborough: Wills and Hepworth.

Southgate, V. 1975. *Beauty and the Beast: A Ladybird Easy-Reading Book.* Loughborough: Ladybird Publishing.

Southgate, V. 1975b. *Goldilocks and the Three Bears: A Ladybird Easy-Reading Book.* Loughborough: Ladybird Publishing.

Southgate, V. 1975c. *Puss-in-Boots: A Ladybird Easy-Reading Book.* Loughborough: Ladybird Publishing.

Spender, J. C. 2008. 'Constructivism'. In R. Thorpe and R. Holt (eds), *The Sage Dictionary of Qualitative Management Research*, pp. 213–215. Los Angeles: Sage Publications.

Stahl, S. K. D. 1977. 'The personal narrative as folklore'. *Journal of the Folklore Institute* 14(1/2): 9–32.

Steenhuizen, D. Flores-Colen, I., Reitsma A. and Lo P. B. 2014. 'The road to facilities management'. *Facilities* 32(1/2): 46–57.

Steig, W. 1990. *Shrek.* New York: Farrar, Straus and Giroux.

Steiner, J. 2006. 'The art of space management'. *Journal of Facilities Management* 4(1): 6–22.

St-James, Y., Handelman, J. M., and Taylor, S. F. 2011. 'Magical thinking and consumer coping'. *Journal of Consumer Research* 38(4): 632–649.

Stiles, D. R. 2004. 'Pictorial Representation'. In C. Cassell and G. Symon (eds), *Essential Guide to Qualitative Methods in Organizational Research*, pp. 127–139. London: Sage Publications.
Stoddart, H. 2007. *Angela Carter's* Nights at the Circus. London: Routledge.
Straker, L., Mathiassen, S. E., and Holtermann, A. 2018. 'The Goldilocks principle: designing physical activity at work to be "just right" for promoting health'. *British Journal of Sports Medicine* 52(13): 818–820.
Strati, A. 1997. 'Organization as hypertext: a metaphor from visual cultures'. *Studies in Cultures, Organizations and Societies* 3: 307–324.
Strati, A. 1998. 'Organizational symbolism as a social construction: a perspective from the sociology of knowledge'. *Human Relations* 51(11): 379–1402.
Styhre, A. 2017. 'Ravaisson, Simondon, and constitution of routine action: organizational routines as habit and individuation'. *Culture and Organization* 23(1): 4–25.
Sveningsson, S., and Alvesson, M. 2003. 'Managing managerial identities: organizational fragmentation, discourse and identity struggle'. *Human Relations* 56(10): 163–1193.
Taleb, N. N. 2007. *The Black Swan: The Impact of the Highly Improbable*. London: Penguin Books.
Tangherlini, T. R. 2000. 'Heroes and lies: storytelling tactics among paramedics'. *Folklore* 111: 43–66.
Taschner, M., and Clayton, J. 2022 Evolution of FM and Review of Canadian FM Practitioner Skills Available at: https://knowledgelibrary.ifma.org/evolution-of-fm-and-review-of-canadian-fm-practitioners-skills/
Tatar, M. 1987. *The Hard Facts of the Grimms' Fairy Tales*. Princeton, NJ: Princeton University Press.
Tatar, M. (ed.) 2002. *The Annotated Classic Fairy Tales*. London: WW Norton and Co.
Tatar, M. (ed.) 2015. *The Cambridge Companion to Fairy Tales*. Cambridge: Cambridge University Press.
Tay, L., and Ooi, J. T. L. 2001. 'Facilities management: a jack-of-all-trades'. *Facilities* 19 (10): 357–363.
Terras, M. 2006. 'Disciplined: using educational studies to analyse humanities computing'. *Literary and Linguistic Computing* 21(2): 229–246.
Teverson, A. 2013. *Fairy Tale*. London: Routledge.
Thanem, T. 2011. *The Monstrous Organization*. Cheltenham: Edward Elgar.
The Statistics Portal 2019. [online] Accessed 16 April 2019 at https://www.statista.com/statistics/321640/news-magazines-ranked-by-sales-volume-uk/
Thompson, S. 1946. *The Folktale*. Berkeley: University of California Press.
Thomson, T. 1990. 'The essence of facilities management'. *Facilities* 8(8): 8–12.
Todorov, T. 1975. *The Fantastic: A Structural Approach to a Literary Genre*. Ithaca, NY: Cornell University Press.
Tolkien, J. R. R. 1947. *On Fairy Stories*. Oxford: Oxford University Press.
Tolovaj Publishing House 2020. 'The history behind the story of Goldilocks'. Accessed 26 March 2021 at https://owlcation.com/humanities/goldilocks-and-three-bears
Toomeos-Orglaan, K. 2013. 'Gender stereotypes in *Cinderella* ATU 510a and *The Princess on the Glass Mountain*, ATU 530'. *Journal of Ethnology and Folkloristics* 7(2): 49–64.
Tranfield, D., and Akhlaghi, F. 1995. 'Performance measures: relating facilities to business indictors'. *Facilities* 13(3): 6–14.
Trice, H. M., and Beyer, J. M. 1984. 'Studying organizational cultures though rites and ceremonials'. *The Academy of Management Review* 9(4): 653–669.
Trubshaw, B. 2002. *Explore Folklore*. Loughborough: Explore Books.
Trubshaw, B. 2003. *Explore Mythology*. Loughborough: Explore Books.

Tsoukas, H. 1991. 'The missing link: a transformational view of metaphors in organizational science'. *The Academy of Management Review* 16(3): 566–585.

Tucker, M., and Masuri, M. 2018. 'The development of facilities management-development process (FM-DP) integration framework'. *Journal of Building Engineering* 18(2): 377–385.

Tyler, J. A. 2011. 'Reclaiming rare listening as a means of organizational re-enchantment'. *Journal of Organizational Change Management* 24(1): 43–157.

Uther, H. 1997. 'Indexing folktales: a critical survey'. *Journal of Folklore Research* 34(3): 209–220.

Uther, H. 2004. 'The types of international folktales: a classification and bibliography'. *FF Communications* 284: 1–16.

van der Geest, S. 2010. 'Rumpelstiltskin: The Magic of the Right Word'. In A. Oderwald, W. van Tilberg, K. Neuvel, and M. Krinkels (eds), *Unfamiliar Knowledge: Psychiatric Disorders in Literature*, pp. 37–44. London: Blue Bird Publishers.

Van Hulst, M., and Ybema, S. 2020. 'From what to where: a setting-sensitive approach to organizational storytelling'. *Organization Studies* 41(3): 365–391.

Van Mannen, J. 1973. 'Observations on the making of policemen'. *Human Organizations* 32(4): 407–418.

Varcoe, B. 1993. 'Facilities systems—what of the future?' *Facilities* 11(10): 7–11.

Varcoe, B., and Hinks, J. 2014a. 'Cost-centric collusion or a customer-centric convergence on integrations. Corporate real estate and facilities management'. [online] Zurich Insurance Group plc. Accessed 16 December 2022 at https://realestateresonance.files.wordpress.com/2014/03/crefm_choice-final.pdf

Varcoe, B., and Hinks, J., 2014b. 'Futures forum 2. Corporate real estate and facilities management'. [online] Zurich Insurance Group plc. Accessed 16 December 2022 at https://realestateresonance.wordpress.com/2012/07/20/crefm-futures-forum/

Van Der Lippe, T., and Lippĕnyi, Z. 2019. 'Co-worker working from home and individual and team performance'. *New Technology, Work and Employment* 35(1): 60–79.

van Marrewijk, A., and Yanow, D. (eds) (2010) *Organizational Spaces: Rematerializing the Workaday World.* Edward Elgar Publishing Limited: Cheltenham. Ventovuori, T., Lehtonen.

Ventovuori, T., Lehtonen, T., Salonen, A., and Nenonen, S. 2007. 'A review and classification of academic research in facilities management'. *Facilities* 25(5/6): 227–237.

Virtanen, L. 1986. 'Modern Folklore: problems of comparative research'. *Journal of Folklore Research* 23(2): 221–232.

Vo, L. C., and Kelemen, M. 2017. 'Collaborating across the researcher–practitioner divide: introducing John Dewey's democratic experimentalism. *Journal of Organizational Change Management* 30(6): 858–871.

Volberda, H. W., Khanagha, S., Baden-Fuller, C., Mihalache, O. R., and Birkinshaw, J. 2021. 'Strategizing in a digital world: overcoming cognitive barriers, reconfiguring routines and introducing new organizational forms'. *Long Range Planning* 54(5): 102–110.

Waheed, Z., and Fernie, S. 2009. 'Knowledge-based facilities management'. *Facilities* 27(70): 258–266.

Wama, T., and Nakatsu R. 2008. 'Analysis and generation of Japanese folktales based on Vladimir Propp's methodology'. In First IEEE International Conference on Ubi-Media Computing 31 July 2008. IEEE.

Warner, M. 1994. *From the Beast to the Blonde: On Fairy Tales and their Tellers.* London: Chatto and Windus.

Warner, M. 2014a. *Once upon a Time: A Short History of Fairy Tale.* Oxford: Oxford University Press.

Warner, M. 2014b. 'The top ten fairytales'. *The Guardian*, 8 October. Accessed 24 March 2022 at http://www.theguardian.com/books/2014/oct/08/top-10-fairytales-marina-warner

Wasserman, V., and Frenkel, M. 2015. 'Spatial work in between glass ceilings and glass walls: gender–class intersectionality and organizational aesthetics'. *Organizational Studies* 36(11): 485–1505.

Watson, C. 2011. 'Staking a small claim for fictional narratives in social and educational research'. *Qualitative Research* 11(4): 395–408.

Watson, T. J. 2000. 'Ethnographic fiction science: making sense of managerial work and organizational research processes with Caroline and Terry'. *Organization* 7(3): 489–510.

Watson, T. J. 2011. 'Ethnography, reality, and truth: the vital need for studies of "how things work" in organizations and management'. *Journal of Management Studies* 48(1): 202–217.

Weber, M. 1930. *The Protestant Ethic and the Spirit of Capitalism.* London: Routledge.

Weber, M. 1947. *Theory of Social and Economic Organisation.* Translated by A. M. Henderson and T. Parsons, originally published 1915. New York: Free Press.

Weber, S. 2008. 'Visual Images in Research'. In J. G. Knowles and A. L. Cole (eds), *The Handbook of the Arts in Qualitative Research*, pp. 41–53. London: Sage Publications.

Weick, K. E. 1995. *Sense-making in Organizations.* London: Sage Publications.

Weick, K. E. 2005. 'Organizing and the process of sense-making'. *Organizational Science* 16(4): 409–421.

Welsch, R. L. 1968. 'A note on definitions'. *Journal of American Folklore* 81: 262–264.

Wengraf, T. 2001. *Qualitative Research Interviewing: Biographic Narrative and Semi-structured Method.* London: Sage Publications.

Westwood, R., Jack, G., Khan, F. R., and Frenkel, M. 2014. *Core-Periphery Relations and Organization Studies.* London: Palgrave.

Whiteman, G., and Philips, N. 2008. 'The Role of Narrative Fiction and Semi-fiction on Organizational Studies'. In D. Barry and H. Hansen (eds), *The Sage Handbook of New Approaches in Management and Organization*, pp. 1–34. London: Sage Publications.

Wiggins, J. M. 2010. *Facilities Manager's Desk Reference.* London: Wiley Blackwell.

Wilcox, L. 1992. 'Making IT strategy succeed: creating a strategic alliance between IT and the business'. *Journal of Information Technology* 7: 57–61.

Wilde, O. 2022. Quote. Accessed 16 December 2022 at http://www.brainyquote.com/quotes/quotes/o/oscarwilde104298.html

Woodside, A., Sood, S., and Miller, K. 2008. 'When consumers and brands talk: storytelling theory and research in psychology and marketing'. *Psychology and Marketing* 25(2): 97–145.

Yang, C. 2013. 'Telling tales at work: an evolutionary explanation'. *Business Communication Quarterly* 76(6): 32–154.

Yeung, C. L., Cheung, C. F., Ming, W., Wang, E., and Lee T. W. B. 2016. 'Managing knowledge in the construction industry through computational generation of semi-fiction narratives'. *Journal of Knowledge Management* 20(2): 386–414.

Zipes, J. 1979. *Breaking the Magic Spell: Radical Theories of Folk and Fairy Tales.* Lexington: University Press of Kentucky.

Zipes, J. 1982. 'The potential of liberating fairy tales for children'. *New Literary History* 13(2): 309–325.

Zipes, J. 1983. *Fairy Tales and the Art of Subversion: The Classical Genre for Children and the Process of Civilisation.* London: Routledge.

Zipes, J. (ed.) 1991. *Spells of Enchantment: The Wondrous Fairy tales of Western Culture.* London: Viking Penguin.

Zipes, J. 1994. *Fairy Tale as Myth, Myth as Fairy tale.* Lexington: University Press of Kentucky.

Zipes, J. 1995. *Creative Storytelling: Building Community and Changing Lives.* London: Routledge.

Zipes, J. 1997. *Happily Ever After: Fairy Tales, Children and the Culture Industry.* London: Routledge.

Zipes, J. (ed.) 2000. *The Oxford Companion to Fairy Tales: The Western Fairy Tales' Tradition from Medieval to Modern.* Oxford: Oxford University Press.

Zipes, J. 2005. 'To eat or not to be eaten: the survival of traditional storytelling'. *Storytelling, Self, Society* 2(1): 1–18.

Zipes, J. 2006. *Why Fairy Tales Stick—The Evolution and Relevance of a Genre.* London: Routledge.

Zipes, J. 2007. *When Dreams Came True: Classical Fairytales and their Tradition*, second edition. London: Routledge.

Zipes, J. 2011. 'The meaning of fairy tales within the evolution of culture'. *Marvels and Tales* 25(2): 221–243.

Zipes, J. (ed.) 2013. *The Golden Age of Folk and Fairy Tales: From the Brothers Grimm to Andrew Lang.* Indianapolis: Hackett Publishing.

Zipes, J. (ed.) 2014a. *The Complete First Edition of the Original Folk and Fairy Tales of the Brothers Grimm.* Princeton: Princeton University Press.

Zipes, J. 2014b. 'Toward a theory of the fairy tales as literary genre'. In Children's Literature Week Stop and Rewind: Remembering the Genre That Shaped Us. 1–7 February. Office of the Dean of Arts and Sciences. https://www.academia.edu/6192587/Fairy_Tale_Theory_Genre

Zipes, J. 2016. 'Once upon a time: changing the world through storytelling'. *Common Knowledge* 22(2): 277–283.

Zundel, M., Holt, R., and Popp, A. 2016. 'Using history in the creation of organizational identity'. *Management and Organizational History* 11(2): 211–235.

Author Index

For the benefit of digital users, indexed terms that span two pages (e.g., 52–53) may, on occasion, appear on only one of those pages.

Note: The following abbreviations have been used – *f* denotes figures or illustrations; *t* denotes tables.

Index

For the benefit of digital users, indexed terms that span two pages (e.g., 52–53) may, on occasion, appear on only one of those pages.

Note: The following abbreviations have been used – *f* denotes figures or illustrations; *t* denotes tables.